Meet MINITAB®

Student Release 14

for Windows®

THOMSON

BROOKS/COLE

Australia • Canada • Mexico • Singapore • Spain
United Kingdom • United States

Printed in Canada
1 2 3 4 5 6 7 08 07 06 05 04

For more information about our products, contact us at:
Thomson Learning Academic Resource Center 1-800-423-0563

For permission to use material from this text or product,
submit a request online at http://www.thomsonrights.com.

Thomson Brooks/Cole
10 Davis Drive
Belmont, CA 94002
USA

Asia
Thomson Learning
5 Shenton Way #01-01
UIC Building
Singapore 068808

Australia/New Zealand
Thomson Learning
102 Dodds Street
Southbank, Victoria 3006
Australia

Canada
Nelson
1120 Birchmount Road
Toronto, Ontario M1K 5G4
Canada

Europe/Middle East/Africa
Thomson Learning
High Holborn House
50/51 Bedford Row
London WC1R 4LR
United Kingdom

Latin America
Thomson Learning
Seneca, 53
Colonia Polanco
11560 Mexico D.F.
Mexico

Spain/Portugal
Paraninfo
Calle Magallanes, 25
28015 Madrid, Spain

ISBN 0-534-41975-5

Table of Contents

Documentation
Additional MINITAB Products
How to Order Additional Products

1
Getting Started

Objectives

In this chapter, you:

- Learn how to use Meet MINITAB, page 1-1
- Start MINITAB, page 1-3
- Open and examine a worksheet, page 1-4

Overview

Meet MINITAB introduces you to the most commonly used features in MINITAB. Throughout the book, you use functions, create graphs, and generate statistics. The contents of *Meet MINITAB* relate to the actions you need to perform in your own MINITAB sessions. You use a sampling of MINITAB's features to see the range of features and statistics that MINITAB provides.

Most statistical analyses require a series of steps, often directed by background knowledge or by the subject area you are investigating. Chapters 2 through 4 illustrate the analysis steps in a typical MINITAB session:

- Exploring data with graphs
- Conducting statistical analyses and procedures
- Assessing quality

Chapters 5 through 8 provide information on:

- Using shortcuts to automate future analyses
- Generating a report
- Preparing worksheets
- Customizing MINITAB to fit your needs

Chapter 9, *Getting Help*, includes information on getting answers and using MINITAB Help features. Chapter 10, *Reference*, provides an overview of the MINITAB environment, a discussion about the types and forms of data that MINITAB uses, and quick-reference tables of actions and statistics available in MINITAB.

You can work through *Meet MINITAB* in two ways:

- From beginning to end, following the story of a fictional online bookstore through a common workflow

- By selecting a specific chapter to familiarize yourself with a particular area of MINITAB

Meet MINITAB introduces dialog boxes and windows when you need them to perform a step in the analysis. As you work, look for these icons for additional information:

    Provides notes and tips

    Suggests related topics in MINITAB Help and StatGuide

Typographical Conventions in this Book

Enter	Denotes a key, such as the **Enter** key.
Alt + D	Denotes holding down the first key and pressing the second key. For example, while holding down the Alt key, press the D key.
File ➤ Exit	Denotes a menu command, in this case choose **Exit** from the **File** menu. Here is another example: **Stat ➤ Tables ➤ Tally Individual Variables** means open the **Stat** menu, then open the **Tables** submenu, and finally choose **Tally Individual Variables**.
Click **OK**.	Bold text clarifies dialog box items and buttons and MINITAB commands.
Enter *Pulse1*.	Italic text specifies text you need to enter.

The Story

An online book retail company has three regional shipping centers that distribute orders to consumers. Each shipping center uses a different computer system to enter and process order information. To integrate all orders and use the most efficient method company wide, the company wants to use the same computer system at all three shipping centers.

Throughout this book, you analyze data from the shipping centers as you learn to use MINITAB. You create graphs and conduct statistical analyses to determine which computer system is the most efficient and results in the shortest delivery time.

After you identify the most efficient computer system, you focus on the data from this center. You create control charts to see whether the center's shipping process is in control.

Additionally, you learn about session commands, generating a report, preparing a worksheet, and customizing MINITAB.

Starting MINITAB

Before you begin your analysis, start MINITAB and examine the layout of the windows.

Start MINITAB   1 From the Windows Taskbar, choose **Start ➤ Programs ➤ MINITAB 14 Student ➤ MINITAB 14 Student**.

MINITAB opens with two main windows visible:

■ The Session window displays the results of your analysis in text format. Also, in this window, you can enter commands instead of using MINITAB's menus.

■ The Data window contains an open worksheet, which is similar in appearance to a spreadsheet. You can open multiple worksheets—each in a different Data window.

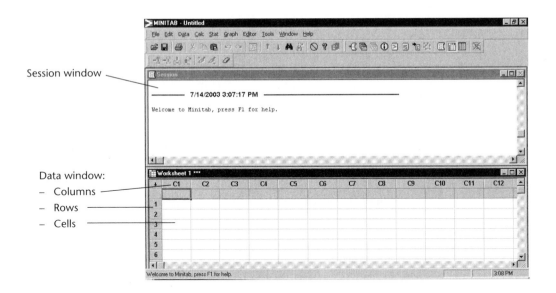

For more information on the MINITAB environment, see *The MINITAB Environment* on page 10-2.

Opening a Worksheet

The data for the three shipping centers are stored in the worksheet SHIPPINGDATA.MTW.

 In some cases, you will need to prepare your worksheet before you begin an analysis. For information on setting up a worksheet, see Chapter 7, *Preparing a Worksheet*.

Open a worksheet

1   Choose **File ➤ Open Worksheet**.

2   In the Data folder, double-click Meet MINITAB.

    You can change the default folder for opening and saving MINITAB files by choosing **Tools ➤ Options ➤ General**.

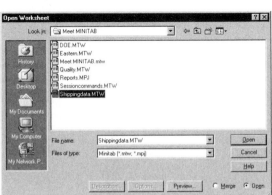

3   Choose SHIPPINGDATA.MTW, then click **Open**. If you get a message box, check **Do not display this message again**, then click **OK**. To restore this message for every

time you open a worksheet, return to MINITAB's default settings. See *Restoring MINITAB's Default Settings* on page 8-6.

Examine worksheet

The data are arranged in columns, which are also called *variables*. The column number and name are at the top of each column. Each row in the worksheet represents a case, which is information on a single book order.

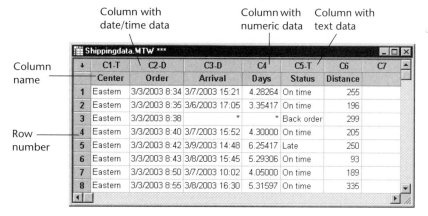

MINITAB accepts three types of data: numeric, text, and date/time. This worksheet contains each type.

The data include:

- Shipping center name

- Order date

- Delivery date

- Number of delivery days

- Delivery status ("On time" indicates that the book shipment was received on time; "Back order" indicates that the book is not currently in stock; "Late" indicates that the book shipment was received six or more days after ordered)

- Distance from shipping center to delivery location

 For more information about data types, see *MINITAB Data* on page 10-5.

What Next

Now that you have a worksheet open, you are ready to start using MINITAB. In the next chapter, you use graphs to check the data for normality and examine the relationships between variables.

2

Graphing Data

Objectives

In this chapter, you:

- Create and interpret an individuals value plot, page 2-2
- Create a histogram with groups, page 2-4
- Edit a histogram, page 2-5
- Arrange multiple histograms on the same page, page 2-6
- Access Help, page 2-8
- Create and interpret scatterplots, page 2-9
- Edit a scatterplot, page 2-10
- Arrange multiple graphs on the same page, page 2-12
- Print graphs, page 2-13
- Save a project, page 2-13

Overview

Before conducting a statistical analysis, you can use graphs to explore data and assess relationships among the variables. Also, graphs are useful to summarize findings and to ease interpretation of statistical results.

You can access MINITAB's graphs from the Graph and Stat menus. Built-in graphs, which help you to interpret results and assess the validity of statistical assumptions, are also available with many statistical commands.

Graph features in MINITAB include:

- A pictorial gallery from which to choose a graph type
- Flexibility in customizing graphs, from subsetting of data to specifying titles and footnotes

- Ability to change most graph elements, such as fonts, symbols, lines, placement of tick marks, and data display, after the graph is created

- Ability to automatically update graphs

This chapter explores the shipping center data you opened in the previous chapter, using graphs to compare means, explore variability, check normality, and examine the relationship between variables.

For more information on MINITAB graphs:

- Go to *Graph overview* in the MINITAB Help index for details on MINITAB graphs. To access the Help index, choose **Help ➤ Help**, then click the **Index** tab.

- Choose **Help ➤ Tutorials ➤ Session One: Graphing Data** for a step-by-step tutorial on using MINITAB graphs and editing tools.

Exploring the Data

Before conducting a statistical analysis, it is a good idea first to create graphs that display important characteristics of the data.

For the shipping center data, you want to know the mean delivery time for each shipping center and how variable the data are within each shipping center. You also want to determine if the shipping center data follow a normal distribution so you that you can use standard statistical methods for testing the equality of means.

Create an individual value plot

You suspect that delivery time is different for the three shipping centers. Create an individual value plot to compare the shipping center data.

1 If not continuing from the previous chapter, choose **File ➤ Open Worksheet**. If continuing from the previous chapter, go to step 3.

2 Double-click Meet MINITAB, then choose SHIPPINGDATA.MTW. Click **Open**.

3 Choose **Graph ➤ Individual Value Plot**.

   For most graphs, MINITAB displays a pictorial gallery. Your gallery choice determines the available graph creation options.

4 Under **One Y**, choose **With Groups**. Click **OK**.

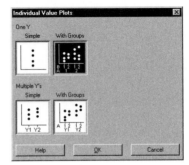

5　In **Graph variables**, enter *Days*.

6　In **Categorical variables for grouping (1-4, outermost first)**, enter *Center*.

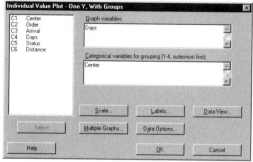

To create a graph, you only need to complete the main dialog box. However, you can click any button to open dialog boxes to customize your graph.

The list box on the left shows the variables from the worksheet that are available for the analysis. The boxes on the right display the variables that you select for the analysis.

7　Click **Data View**. Check **Mean connect line**.

8　Click **OK** in each dialog box.

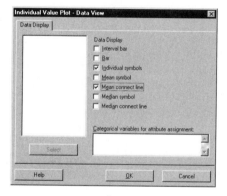

 To select variables in most MINITAB dialog boxes, you can: double-click the variables in the variables list box; highlight the variables in the list box, then choose **Select**; or type the variables' names or column numbers. For more information, go to *Dialog boxes, Selecting variables* in the MINITAB Help index.

Graph window output

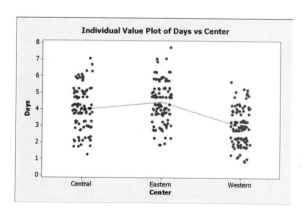

Interpret results

The individual value plots show that each center has a different mean delivery time. The Western center has a lower shipping time than the Central and Eastern centers. The variation within each shipping center seems about the same.

Create a grouped histogram

Another way to compare the three shipping centers is to create a grouped histogram, which displays the histograms for each center on the same graph. The grouped histogram will show how much the data from each shipping center overlap.

1 Choose **Graph ➤ Histogram**.

2 Choose **With Fit And Groups**, then click **OK**.

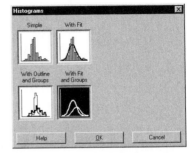

3 In **Graph variables**, enter *Days*.

4 In **Categorical variables for grouping (0-3)**, enter *Center*.

5 Click **OK**.

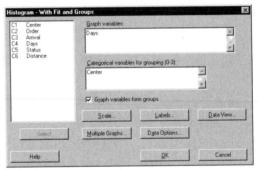

Graph window output

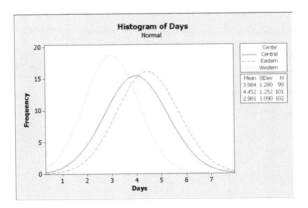

Interpret results

As you saw in the individual value plot, the means for each center are different. The mean delivery times are:

Central — 3.984 days

Eastern — 4.452 days

Western — 2.981 days

The grouped histogram shows that the Central and Eastern centers are similar in mean delivery time and spread of delivery time. In contrast, the Western center mean delivery time is shorter and less spread out. Chapter 3, *Analyzing Data*, shows how to detect stastistically significant differences among means using analysis of variance.

If your data change, MINITAB can automatically update graphs. For more information, go to *Update (Editor menu)* in the MINITAB Help index.

Edit histogram

Editing graphs in MINITAB is easy. You can edit virtually any graph element. For the histogram you just created, you want to:

- Increase the size of the text in the legend (the table with the center information) and the table that contains the Mean, StDev, and N

- Modify the title

Change the output table font

1 Double-click the legend.

2 Click the **Font** tab.

3 Under **Size**, choose *10*.

4 Click **OK**.

5 Repeat steps 1–4 for the table.

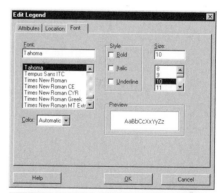

Change the title

1 Double-click the title (*Histogram of Days*).

2 In **Text**, type *Histogram of Delivery Time*.

3 Click **OK**.

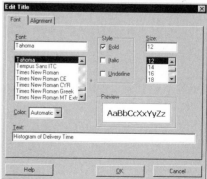

Graph window output

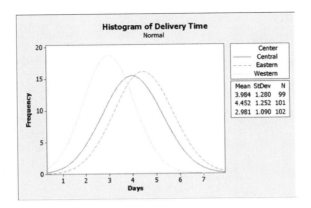

Interpret results

The histogram now features a larger font for the output table and a more descriptive title.

In addition to editing individual graphs, you can change the default settings for future graphs.

- To affect general graph settings, such as font attributes, graph size, and line types, choose **Tools ➤ Options ➤ Graphics**.
- To affect graph-specific settings, such as the scale type on histograms or the method for calculating the plotted points on probability plots, choose **Tools ➤ Options ➤ Individual Graphs**.

The next time you open an affected dialog box, your preferences are reflected.

Create a paneled histogram

To determine if the shipping center data follow a normal distribution, create a paneled histogram of the time lapse between order and delivery date.

1 Choose **Graph ➤ Histogram**.

2 Choose **With Fit**, then click **OK**.

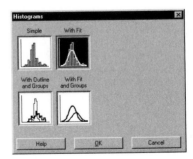

3 In **Graph variables**, enter *Days*.

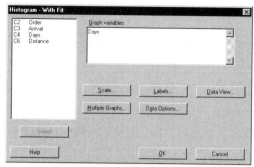

4 Click **Multiple Graphs**, then click the **By Variables** tab.

5 In **By variables with groups in separate panels**, enter *Center*.

6 Click **OK** in each dialog box.

Graph window output

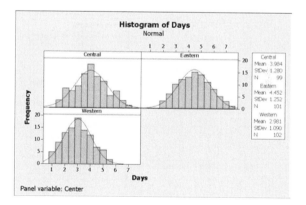

Interpret results

The delivery times for each center are approximately normally distributed as shown by the distribution curves exhibiting the same pattern.

If you have fewer than 50 observations, you may want to use a normal probability plot (**Graph ➤ Probability Plot** or **Stat ➤ Basic Statistics ➤ Normality Test**) to assess normality.

Examining Relationships Between Two Variables

Graphs can help identify whether associations are present among variables and the strength of any associations. Knowing the relationship among variables can help to guide further analyses and determine which variables are important to analyze.

Because each shipping center serves a small regional delivery area, you suspect that distance to delivery site does not greatly affect delivery time. To verify this suspicion and eliminate distance as a potentially important factor, examine the relationship between delivery time and delivery distance.

Access Help

To find out which graph shows the relationship between two variables, use MINITAB Help.

1 Choose **Help ➤ Help**.

2 Click the **Index** tab.

3 In **Type in the keyword to find**, type *Graph overview.*

4 Double-click the **Graph overview** index entry to access the Help topic.

5 In the Help topic, under the heading **Types of graphs**, click **Examine relationships between pairs of variables**.

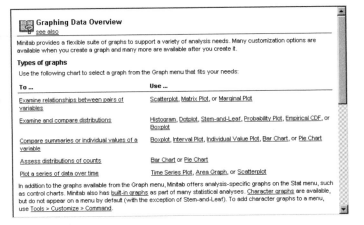

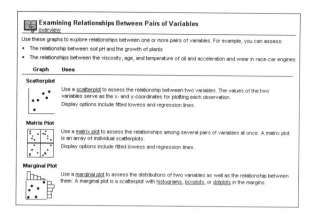

This Help topic suggests that a scatterplot is the best choice to see the relationship between delivery time and delivery distance.

 For help on any MINITAB dialog box, click **Help** in the lower left corner of the dialog box or press F1 . For more information on MINITAB Help, see Chapter 9, *Getting Help*.

Create a scatterplot

1 Choose **Graph ➤ Scatterplot**.

2 Choose **With Regression**, then click **OK**.

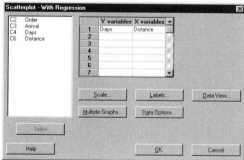

3 Under **Y variables**, enter *Days*.
   Under **X variables**, enter *Distance*.

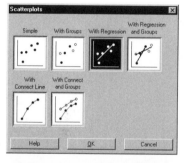

4 Click **Multiple Graphs**, then click the **By Variables** tab.

5 In **By variables with groups in separate panels**, enter *Center*.

6 Click **OK** in each dialog box.

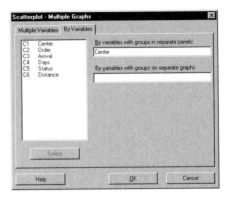

Graph window output

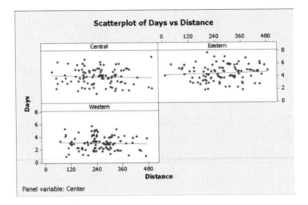

Interpret results

The points on the scatterplot exhibit no apparent pattern at any of the three centers. The regression line for each center is relatively flat, suggesting that the proximity of a delivery location to a shipping center does not affect the delivery time.

Edit scatterplot

To help your colleagues quickly interpret the scatterplot, you want to add a footnote to the plot.

1 Click the scatterplot to make it active.

2 Choose **Editor ➤ Add ➤ Footnote**.

3 In **Footnote**, type *Relationship between delivery time and distance from shipping center*.

4 Click **OK**.

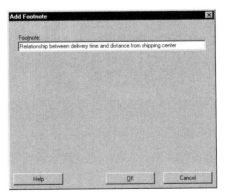

*Graph
window
output*

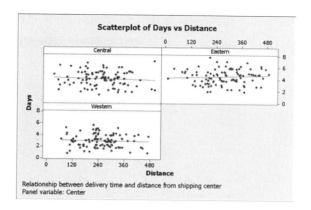

**Interpret
results**

The scatterplot now features a footnote that provides a brief interpretation of the results.

Using Graph Layout and Printing

Use MINITAB's graph layout tool to place multiple graphs on the same page. You can add annotations to the layout and edit the individual graphs within the layout.

To show your supervisor the preliminary results of the graphical analysis of the shipping data, display all four graphs on one page.

When you issue a MINITAB command that you previously used in the same session, MINITAB remembers the dialog box settings. To set a dialog box back to its defaults, press F3 .

Create graph layout

1 With the scatterplot active, choose **Editor ➤ Layout Tool**. The active graph, the scatterplot, is already included in the layout.

A list of all open graphs

Buttons used to move graphs to and from the layout

The next graph to be moved to the layout

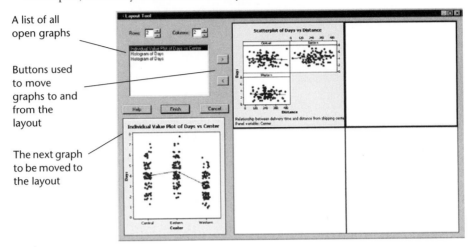

2 Click the scatterplot and drag it to the bottom right corner of the layout.

3 Click ⧁ to place the individual value plot in the upper-left corner of the layout.

4 Click ⧁ to place the grouped histogram in the upper-right corner.

5 Click ⧁ to place the paneled histogram in the lower-left corner.

6 Click **Finish**.

Graph window output

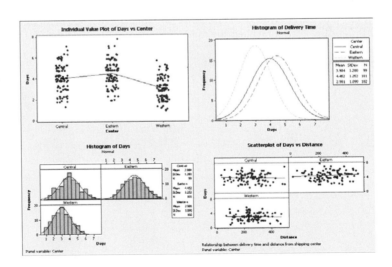

If the worksheet data change after you create a layout, MINITAB does not automatically update the graphs in the layout. You must re-create the layout with the updated individual graphs.

Annotate the layout

You want to add a descriptive title to the layout.

1 Choose **Editor** ➤ **Add** ➤ **Title**.

2 In **Title**, type *Graphical Analysis of Shipping Center Data*. Click **OK**.

Graph window output

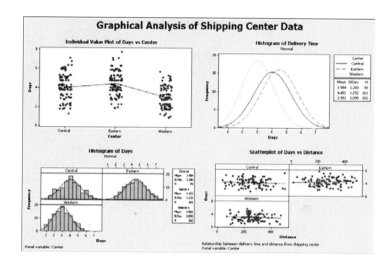

Print graph layout

You can print an individual graph or a layout just as you would any other MINITAB window.

1 Click the Graph window to make it active, then choose **File** ➤ **Print Graph**.

2 Click **OK**.

Saving Projects

MINITAB data are saved in worksheets. You can also save MINITAB projects which can contain multiple worksheets. A MINITAB project contains all your work, including the data, Session window output, graphs, history of your session, ReportPad contents, and dialog box settings. When you open a project, you can resume working where you left off.

Save a
MINITAB
project

Save all of your work in a MINITAB project.

1 Choose **File ➤ Save Project As**.

2 In **File name**, type *MY_GRAPHS.MPJ*. MINITAB automatically adds the extension .MPJ to the file name when you save the project.

3 Click **Save**.

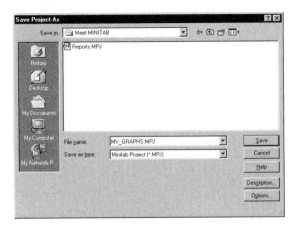

  If you close a project before saving it, MINITAB prompts you to save the project.

What Next

The graphical output indicates that the three shipping centers have different delivery times for book orders. In the next chapter, you display descriptive statistics and perform an analysis of variance (ANOVA) to test whether the differences among the shipping centers are statistically significant.

3
Analyzing Data

Objectives

In this chapter, you:

- Display and interpret descriptive statistics, page 3-2
- Perform and interpret a one-way ANOVA, page 3-4
- Display and interpret built-in graphs, page 3-4
- Access the StatGuide, page 3-8
- Use the Project Manager, page 3-10

Overview

The field of statistics provides principles and methodologies for collecting, summarizing, analyzing, and interpreting data, and for drawing conclusions from analysis results. Statistics can be used to describe data and to make inferences, both of which can guide decisions and improve processes and products.

MINITAB provides:

- Many statistical methods organized by category, such as regression, ANOVA, quality tools, and time series
- Built-in graphs to help you understand the data and validate results
- The ability to display and store statistics and diagnostic measures

This chapter introduces MINITAB's statistical commands, built-in graphs, StatGuide, and Project Manager. You want to assess the number of late and back orders, and test whether the difference in delivery time among the three shipping centers is statistically significant.

 For more information on MINITAB's statistical features, go to *Stat menu* in the MINITAB Help index.

Displaying Descriptive Statistics

Descriptive statistics summarize and describe the prominent features of data.

Use Display Descriptive Statistics to find out how many book orders were delivered on time, how many were late, and the number that were initially back ordered for each shipping center.

Display descriptive statistics

1 If continuing from the previous chapter, choose **File ➤ New**, then choose **Minitab Project**. Click **OK**. Otherwise, just start MINITAB.

2 Choose **File ➤ Open Worksheet**.

3 Double-click Meet MINITAB, then choose SHIPPINGDATA.MTW. Click **Open**. This worksheet is the same one you used in Chapter 2, *Graphing Data*.

4 Choose **Stat ➤ Basic Statistics ➤ Display Descriptive Statistics**.

5 In **Variables**, enter *Days*.

6 In **By variables (optional)**, enter *Center Status*.

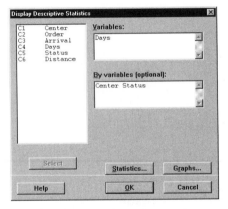

For most MINITAB commands, you only need to complete the main dialog box to execute the command. But, you can often use subdialog boxes to modify the analysis or display additional output, like graphs.

7 Click **Statistics**.

8 Uncheck **First quartile, Median, Third quartile, N nonmissing**, and **N missing**.

9 Check **N total**.

10 Click **OK** in each dialog box.

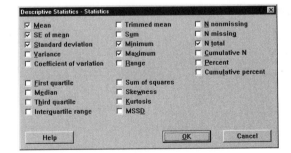

 Changes made in the Statistics subdialog box affect the current session only. To change the default settings for future sessions, use **Tools ➤ Options ➤ Individual Commands ➤ Display Descriptive Statistics**. When you open the Statistics subdialog box again, it reflects your preferences.

*Session
window
output*

Descriptive Statistics: Days

Results for Center = Central

Variable	Status	Total Count	Mean	SE Mean	StDev	Minimum	Maximum
Days	Back order	6	*	*	*	*	*
	Late	6	6.431	0.157	0.385	6.078	7.070
	On time	93	3.826	0.119	1.149	1.267	5.983

Results for Center = Eastern

Variable	Status	Total Count	Mean	SE Mean	StDev	Minimum	Maximum
Days	Back order	8	*	*	*	*	*
	Late	9	6.678	0.180	0.541	6.254	7.748
	On time	92	4.234	0.112	1.077	1.860	5.953

Results for Center = Western

Variable	Status	Total Count	Mean	SE Mean	StDev	Minimum	Maximum
Days	Back order	3	*	*	*	*	*
	On time	102	2.981	0.108	1.090	0.871	5.681

The Session window displays text output, which you can edit, add to the ReportPad, and print. The ReportPad is discussed in Chapter 6, *Generating a Report*.

**Interpret
results**

The Session window presents each center's results separately. Within each center, you can find the number of back, late, and on-time orders in the Total Count column.

- The Eastern shipping center has the most back orders (8) and late orders (9).

- The Central shipping center has the next greatest number of back orders (6) and late orders (6).

- The Western shipping center has the smallest number of back orders (3) and no late orders.

You can also review the Session window output for the mean, standard error of the mean, standard deviation, minimum, and maximum of order status for each center. These statistics are not given for back orders because no delivery information exists for these orders.

Performing an ANOVA

One of the most commonly used methods in statistical decisions is hypothesis testing. MINITAB offers many hypothesis testing options, including t-tests and analysis of variance. Generally, a hypothesis test assumes an initial claim to be true, then tests this claim using sample data.

Hypothesis tests include two hypotheses: the null hypothesis (denoted by H_0) and the alternative hypothesis (denoted by H_1). The null hypothesis is the initial claim and is often specified using previous research or common knowledge. The alternative hypothesis is what you may believe to be true.

Based on the graphical analysis you performed in the previous chapter and the descriptive analysis above, you suspect that difference in the average number of delivery days (response) across shipping centers (factor) is statistically significant. To verify this, perform a one-way ANOVA, which tests the equality of two or more means categorized by a single factor. Also, conduct a Tukey's multiple comparison test to see which shipping center means are different.

Perform an ANOVA

1 Choose **Stat ➤ ANOVA ➤ One-Way**.

2 In **Response**, enter *Days*. In **Factor**, enter *Center*.

   In many dialog boxes for statistical commands, you can choose frequently used or required options. Use the subdialog box buttons to choose other options.

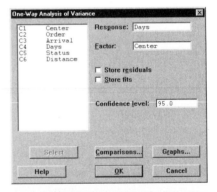

3 Click **Comparisons**.

4 Check **Tukey's, family error rate**, then click **OK**.

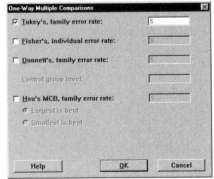

5 Click **Graphs**.

For many statistical commands, MINITAB includes built-in graphs that help you interpret the results and assess the validity of statistical assumptions.

6 Check **Individual value plot** and **Boxplots of data**.

7 Under **Residual Plots**, choose **Four in one**.

8 Click **OK** in each dialog box.

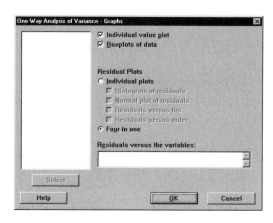

Session window output

One-way ANOVA: Days versus Center

```
Source   DF      SS     MS      F      P
Center    2  114.63  57.32  39.19  0.000
Error   299  437.28   1.46
Total   301  551.92


S = 1.209   R-Sq = 20.77%   R-Sq(adj) = 20.24%
```

```
                                Individual 95% CIs For Mean Based on
                                Pooled StDev
Level      N   Mean  StDev   -----+---------+---------+---------+----
Central   99  3.984  1.280                        (----*---)
Eastern  101  4.452  1.252                                  (----*----)
Western  102  2.981  1.090   (----*---)
                             -----+---------+---------+---------+----
                              3.00      3.50      4.00      4.50


Pooled StDev = 1.209



Tukey 95% Simultaneous Confidence Intervals
All Pairwise Comparisons among Levels of Center

Individual confidence level = 98.01%


Center = Central subtracted from:

Center   Lower  Center   Upper   ---------+---------+---------+---------+
Eastern  0.068   0.468   0.868                       (---*---)
Western -1.402  -1.003  -0.603        (---*---)
                                 ---------+---------+---------+---------+
                                      -1.0       0.0       1.0       2.0
```

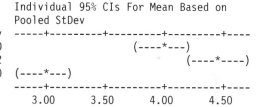

```
Center = Eastern subtracted from:

Center    Lower   Center   Upper   ---------+---------+---------+---------+
Western   -1.868  -1.471   -1.073  (---*---)
                                   ---------+---------+---------+---------+
                                          -1.0      0.0       1.0       2.0
```

Interpret results

The decision-making process for a hypothesis test can be based on the probability value (p-value) for the given test.

- If the p-value is less than or equal to a predetermined level of significance (α-level), then you reject the null hypothesis and claim support for the alternative hypothesis.

- If the p-value is greater than the α-level, you fail to reject the null hypothesis and can not claim support for the alternative hypothesis.

In the ANOVA table, the p-value (0.000) provides sufficient evidence that the average delivery time is different for at least one of the shipping centers from the others when α is 0.05. In the individual 95% confidence intervals table, notice that none of the intervals overlap, which supports the theory that the means are statistically different. However, you need to interpret the multiple comparison results to see where the differences exist among the shipping center averages.

Tukey's test provide two sets of multiple comparison intervals:

- Central shipping center mean subtracted from Eastern and Western shipping center means

- Eastern shipping center mean subtracted from Western center mean

The first interval in the first set of the Tukey output is 0.068 to 0.868. That is, the mean delivery time of the Eastern center minus that of the Central center is somewhere between 0.068 and 0.868 days. Because the interval does not include zero, the difference in delivery time between the two centers is statistically significant. The Eastern center's deliveries take longer than the Central center's deliveries. You similarly interpret the other Tukey test results. The means for all shipping centers differ significantly because all of the confidence intervals exclude zero. Therefore, all the shipping centers have significantly different average delivery times. The Western shipping center has the fastest mean delivery time (2.981 days).

*Graph
window
output*

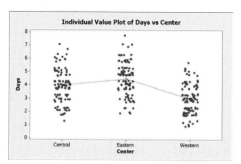

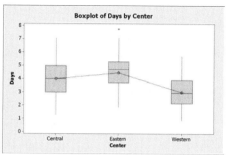

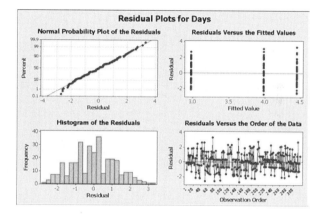

**Interpret
results**

The individual value plots and boxplots indicate that the delivery time varies by shipping center, which is consistent with the graphs from the previous chapter. The boxplot for the Eastern shipping center indicates the presence of one outlier (indicated by ∗), which is an order with an unusually long delivery time.

Use residual plots, available with many statistical commands, to check statistical assumptions:

- Normal probability plot—to detect nonnormality. An approximately straight line indicates that the residuals are normally distributed.

- Histogram of the residuals—to detect multiple peaks, outliers, and nonnormality. The histogram should be approximately symmetric and bell-shaped.

- Residuals versus the fitted values—to detect nonconstant variance, missing higher-order terms, and outliers. The residuals should be scattered randomly around zero.

- Residuals versus order—to detect time-dependence of residuals. The residuals should exhibit no clear pattern.

For the shipping data, the four-in-one residual plots indicate no violations of statistical assumptions. The one-way ANOVA model fits the data reasonably well.

 In MINITAB, you can display each of the residual plots on a separate page. You can also create a plot of the residuals versus the variables.

Access StatGuide

You want more information on how to interpret a one-way ANOVA, particularly Tukey's multiple comparison test. MINITAB StatGuide provides detailed information about the Session and Graph window output for most statistical commands.

1 Place your cursor anywhere in the one-way ANOVA Session window output.

2 Click 📠 on the Standard toolbar.

The MiniGuide window contains a list of one-way ANOVA topics.

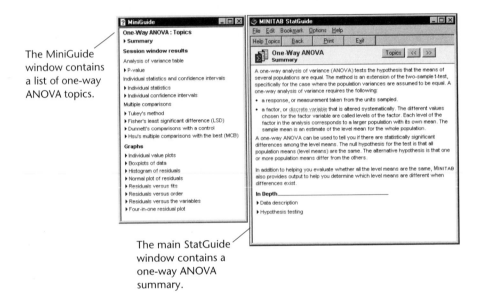

The main StatGuide window contains a one-way ANOVA summary.

3 You want to learn more about Tukey's multiple comparison method. In the MiniGuide window, click **Tukey's method**.

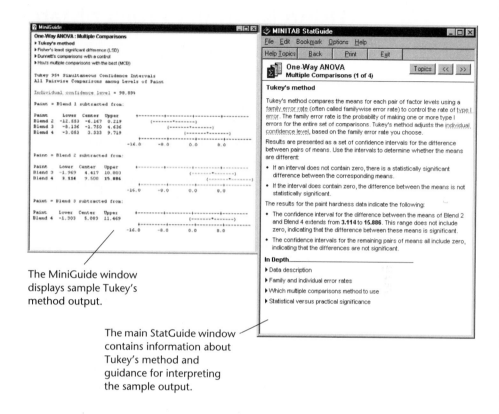

The MiniGuide window displays sample Tukey's method output.

The main StatGuide window contains information about Tukey's method and guidance for interpreting the sample output.

4 If you like, use [<<] [>>] to browse through the one-way ANOVA topics.

5 In the StatGuide window, choose **File ➤ Exit**.

 For more information about using the StatGuide, see *StatGuide* on page 9-8 or choose **Help ➤ How to Use the StatGuide**.

Save project Save all your work in a MINITAB project.

1 Choose **File ➤ Save Project As**.

2 In **File name**, type *MY_STATS.MPJ*.

3 Click **Save**.

Using MINITAB's Project Manager

Now you have a MINITAB project that contains a worksheet, several graphs and Session window output from your analyses. The Project Manager helps you navigate, view, and manipulate parts of your MINITAB project.

Use the Project Manager to view the statistical analyses you just conducted.

Open Project Manager

1 To access the Project Manager, click 🔲 on the Project Manager toolbar or press Ctrl+I.

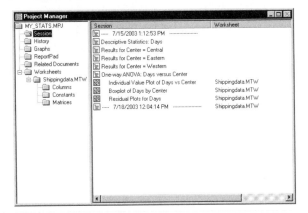

You can easily view the Session window output and graphs by choosing from the list in the right pane. You can also use the icons on the Project Manager toolbar to access different output.

For more information, see *Project Manager* on page 10-3.

View Session window output

You want to review the one-way ANOVA output. To become familiar with the Project Manager toolbar, use the Show Session Folder icon 🔳 on the toolbar, which opens the Session window.

1 Click 🔳 on the Project Manager toolbar.

2 Double-click **One-way ANOVA: Days versus Center** in the left pane.

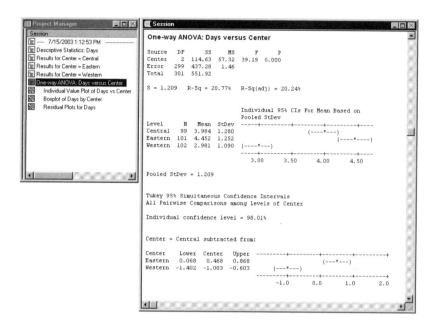

The Project Manager displays the one-way ANOVA Session window output in the right pane.

View graphs

You also want to view the boxplot again. Use the Show Graphs icon 🖼 on the toolbar.

1 Click 🖼 on the Project Manager toolbar.

2 In the left pane, double-click **Boxplot of Days by Center** in the left pane.

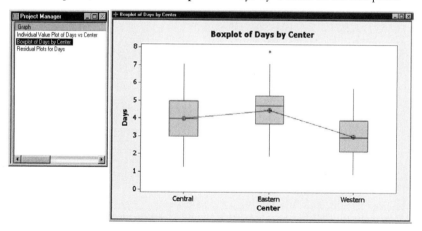

The Project Manager displays the boxplot in the Graph window in the right pane.

What Next

The descriptive statistics and ANOVA results indicate that the Western center has the fewest late and back orders and the shortest delivery time. In the next chapter, you create a control chart to investigate whether the Western shipping center's process is stable over time.

4

Assessing Quality

Objectives

In this chapter, you:

- Set options for control charts, page 4-2

- Create and interpret control charts, page 4-3

- Update a control chart, page 4-5

- View subgroup information, page 4-7

- Add a reference line to a control chart, page 4-7

Overview

Quality is the degree to which products or services meet the needs of customers. Common objectives for quality professionals include reducing defect rates, manufacturing products within specifications, and standardizing delivery time.

MINITAB offers a wide array of methods to help you evaluate quality in an objective, quantitative way: control charts and quality planning tools. This chapter discusses control charts.

Features of MINITAB control charts include:

- The ability to choose how to estimate parameters and control limits, as well as display tests for special causes and historical stages.

- Customizable attributes, such as adding a reference line, changing the scale, and modifying titles. As with other MINITAB graphs, you can customize control charts when and after you create them.

The graphical and statistical analyses conducted in the previous chapter show that the Western shipping center has the fastest delivery time. In this chapter, you determine whether the center's process is stable (in control).

Evaluating Process Stability

Use control charts to track process stability over time and to detect the presence of special causes, which are unusual occurrences that are not a normal part of the process.

MINITAB plots a process statistic—such as a subgroup mean, individual observation, weighted statistic, or number of defects—versus a sample number or time. MINITAB draws the:

- Center line at the average of the statistic
- Upper control limit (UCL) at 3 standard deviations above the center line
- Lower control limit (LCL) at 3 standard deviations below the center line

For all control charts, you can modify MINITAB's default chart specifications. For example, you can define the estimation method for the process standard deviation, specify the tests for special causes, and display process stages by defining historical stages.

 For additional information on MINITAB's control charts, go to *Control Charts (Stat menu)* in the MINITAB Help index.

Set options for control charts

Before you create a control chart for the book shipping data, you want to specify options different from MINITAB's defaults for testing the randomness of the data for all control charts.

The Automotive Industry Action Group (AIAG) suggests using the following guidelines to test for special causes:

- Test 1: 1 point > 3 standard deviations from center line
- Test 2: 9 points in a row on the same side of center line
- Test 3: 6 points in a row, all increasing or all decreasing

Also, in accordance with AIAG guidelines, for all future control charts, you want to use a value of 7 for tests 2 and 3. You can easily do this by setting options for your control charts analysis. When you set options, affected dialog boxes automatically reflect your preferences.

1 Choose **Tools ➤ Options ➤ Control Charts and Quality Tools ➤ Define Tests**.

2 Under **K** for **Test 2**, change the value to 7.

3 Under **K** for **Test 3**, change the value to 7.

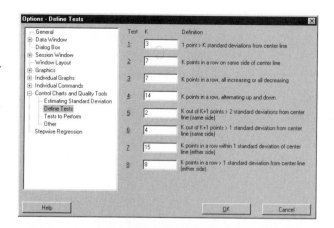

4 Choose **Tests to Perform** in the left pane.

5 Check the first three tests. Notice the values you changed in steps 2 and 3 are reflected in this dialog box.

6 Click **OK**.

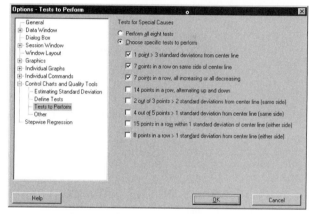

 If you set options, you can restore MINITAB's default settings at any time. For more information, see *Restoring MINITAB's Default Settings* on page 8-6.

Create X̄ and S chart

Now you are ready to create a control chart to see whether the delivery process is stable over time. You randomly select 10 samples for 20 days to examine changes in the mean and variability of delivery time. Create an X̄ and S chart with which you can monitor the process mean and variability simultaneously. Use X̄ and S charts when you have subgroups of size 9 or more.

1 If continuing from the previous chapter, choose **File ➤ New**, then choose **Minitab Project**. Click **OK**. Otherwise, just start MINITAB.

2 Choose **File ➤ Open Worksheet**.

3 Double-click Meet MINITAB, then choose QUALITY.MTW. Click **Open**.

4 Choose **Stat ➤ Control Charts ➤ Variables Charts for Subgroups ➤ Xbar-S**.

To create a control chart, you only need to complete the main dialog box. However, you can click any button to select options for customizing your chart.

5 Choose **All observations for a chart are in one column**, then enter *Days*.

6 In **Subgroup sizes**, enter *Date*.

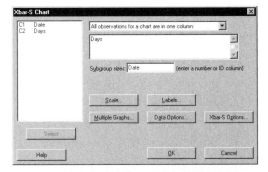

7 Click **Xbar-S Options**, then click the **Tests** tab. Notice this dialog box reflects the tests and test values you specified earlier. (See *Set options for control charts* on page 4-2.)

You can click any tab to open dialog boxes to customize your control chart. Available tabs depend on whatever is appropriate for the chart type. Parameters, Estimate, S Limits, Stages, Display, and Storage are available for all control charts. Tests are available for most charts.

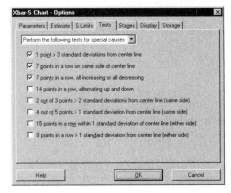

8 Click **OK** in each dialog box.

Graph window output

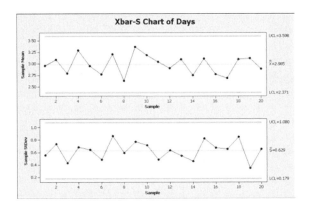

Interpret X̄ and S chart

The data points for the Western shipping center fall within the bounds of the control limits, and do not display any nonrandom patterns. Therefore, the process mean and process standard deviation appear to be in control (stable). The mean ($\overline{\overline{X}}$), is 2.985, and the average standard deviation($\overline{S}$) is 0.629.

Update control chart

Graph updating allows you to update a graph when the data change without re-creating the graph. Graph updating is available for all graphs in the Graph menu (except Stem-and-Leaf) and all control charts.

After creating the $\overline{X}$ and S chart, the Western shipping center manager gives you more data collected on 3/23/2003. Add the data to the worksheet and update the control chart.

Add the data to the worksheet

You need to add both date/time data to C1 and numeric data to C2.

1 Click the Data window to make it active.

2 Place your cursor in any cell in C1, then press End to go to the bottom of the worksheet.

3 To add the date 3/23/2003 to rows 201–210:

- First, type 3/23/2003 in row 201 in C1.

- Then, select the cell containing 3/23/2003, place the cursor over the Autofill handle in the lower-right corner of the highlighted cell. When the mouse is over the handle, a cross symbol (+) appears. Press Ctrl and drag the cursor to row 210 to fill the cells with the repeated date value. When you hold Ctrl down, a superscript cross appears above the Autofill cross symbol (+⁺), indicating that repeated, rather than sequential, values will be added to the cells.

↓	C1-D	C2	C3
	Date	Days	
195	3/22/2003	2.50	
196	3/22/2003	2.85	
197	3/22/2003	2.69	
198	3/22/2003	1.83	
199	3/22/2003	3.59	
200	3/22/2003	2.82	
201	3/23/2003		
202			
203			
204			

Quality.MTW ***

4 Add the following data to C2, starting in row 201:

3.60 2.40 2.80 3.21 2.40 2.75 2.79 3.40 2.58 2.50

If the data entry arrow is facing downward, pressing [Enter] moves the cursor to the next cell down.

Data entry arrow ———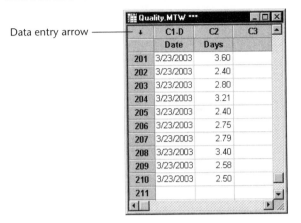

5 Verify that you entered the data correctly.

Update the control chart

1 Right-click the $\overline{X}$ and S chart and choose **Update Graph Now**.

Graph window output

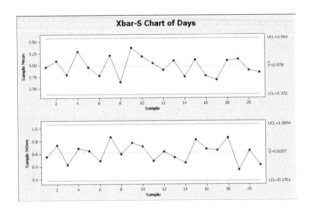

The $\overline{X}$ and S chart now includes the new subgroup. The mean ($\overline{\overline{X}}$ = 2.978) and standard deviation ($\overline{S}$ = 0.6207) have changed slightly, but the process still appears to be in control.

To update all graphs and control charts automatically:
1 Choose **Tools ➤ Options ➤ Graphics ➤ Other Graphics Options**.
2 Check **On creation, set graph to update automatically when data change**.

**View
subgroup
information**

As with any MINITAB graph, when you move your mouse over the points in a control chart, you see various information about the data.

You want to find out the mean of sample 9, the subgroup with the largest mean.

1   Move your mouse over the data point for sample 9.

*Graph
window
output*

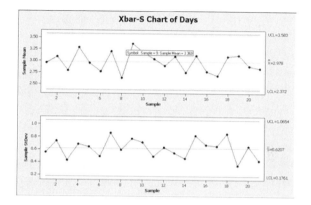

**Interpret
results**

The data tip shows that sample 9 has a mean delivery time of 3.369 days.

**Add
reference line**

A goal for the online bookstore is for all customers to receive their orders in 3.33 days (80 hours) on average, so you want to compare the average delivery time for the Western shipping center to this target. You can show the target level on the $\bar{X}$ chart by adding a reference line.

1   Right-click the $\bar{X}$ chart (the top chart), and choose **Add ▶ Reference Lines**.

2   In **Show reference lines for Y positions**, type 3.33.

3   Click **OK**.

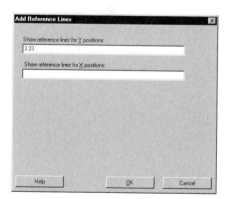

*Graph
window
output*

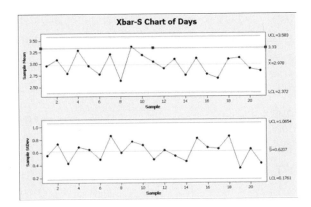

**Interpret
results**

The center line ($\overline{\overline{X}}$) is well below the reference line, indicating that, on average, the Western shipping center delivers books faster than the target of 3.33 days. Only subgroup 9 has a delivery time that falls above the reference line (> 3.33).

1 Save project

Save all of your work in a MINITAB project.

1 Choose **File** ➤ **Save Project As**.

2 In **File name**, type *MY_QUALITY.MPJ*.

3 Click **Save**.

What Next

The quality analysis indicates that the Western shipping center's process is in control. In the next chapter, you learn how to use command language and create and run Execs to quickly rerun an analysis when new data are collected.

5

Using Session Commands

Objectives

In this chapter, you:

- Enable and type session commands, page 5-2

- Conduct an analysis using session commands, page 5-3

- Rerun a series of session commands with Command Line Editor, page 5-5

- Create and run an Exec, page 5-6

Overview

Each menu command has a corresponding session command. Session commands consist of a main command and, in most cases, one or more subcommands. Commands are usually easy-to-remember words, such as PLOT, CHART, or SORT. Both main commands and subcommands can be followed by a series of arguments, which can be columns, constants, or matrices, text strings, or numbers.

Session commands can be:

- Typed into the Session window or the Command Line Editor.

- Copied from the History folder to the Command Line Editor. (When you use menu commands, MINITAB generates and stores the corresponding session commands in the History folder.)

- Copied and saved in a file called an Exec, which can be reexecuted and shared with others or used in future sessions.

Use session commands to quickly rerun an analysis in current or future sessions or as an alternative to menu commands. Some users find session commands quicker to use than menu commands once they become familiar with them.

The Western shipping center continuously collects and analyzes shipping time when new data are available. In Chapter 4, *Assessing Quality*, you created a control chart on data from March. In this chapter, you create a control chart on data from April using session commands.

To learn more about session commands, choose **Help ➤ Session Command Help**.

Enabling and Typing Commands

One way to use session commands is to directly type the commands and subcommands at the command prompt in the Session window. However, MINITAB does not display the command prompt by default. To enter commands directly into the Session window, you must enable this prompt.

Enable session commands

1 If continuing from the previous chapter, choose **File ➤ New**, then choose **Minitab Project** and click **OK**. Otherwise, just start MINITAB.

2 Choose **File ➤ Open Worksheet**.

3 Double-click Meet MINITAB, then choose SESSIONCOMMANDS.MTW. Click **Open**.

4 Click the Session window to make it active.

5 Choose **Editor ➤ Enable Commands**. A check appears next to the menu item.

To change the default options and enable session commands for all future sessions:
1 Choose **Tools ➤ Options ➤ Session Window ➤ Submitting Commands**.
2 Under **Command Language**, click **Enable**.

Examine Session window

With the command prompt enabled, you can now type session commands in the Session window.

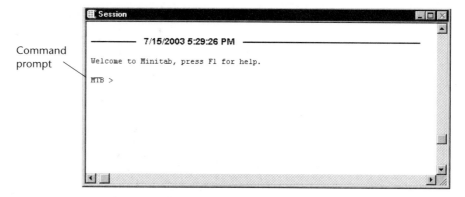

Command prompt

When you execute a command from a menu and session commands are enabled, the corresponding session command appears in the Session window along with your text output. This technique provides a convenient way to learn session commands.

Conduct an analysis with session commands

In Chapter 4, *Assessing Quality*, you created an $\overline{X}$ and S chart to determine whether shipping times were stable for the month of March. To perform this analysis, you used **Stat ➤ Control Charts ➤ Variables Charts for Subgroups ➤ Xbar-S**.

To continue evaluating shipping times at the Western shipping center, you plan to repeat this analysis at regular intervals. When you collect new data, you can re-create this chart using just a few session commands, instead of filling out dialog boxes. Analyze the April shipping data using session commands. Create a new title to indicate that this data is from April.

1 In the Session window, at the MTB > prompt, type:

XSCHART 'Days' 'Date';

2 Press Enter.

The semicolon indicates that you want to type a subcommand.

Notice that the MTB > prompt becomes SUBC>, allowing you to add subcommands for the various options used in the earlier capability analysis.

Subcommand prompt ———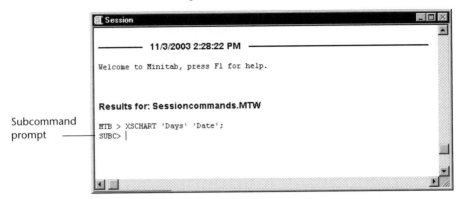

3 At the SUBC> prompt, type:

TITLE "Xbar-S Chart for April Shipping Data".

4 Press [Enter].

The period indicates the end of a command sequence.

MINITAB displays the $\overline{X}$ and S chart for the April shipping data.

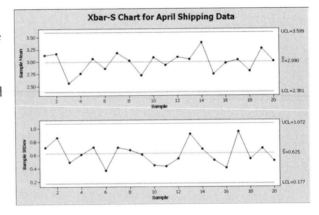

For more information on session commands, including command and subcommand syntax, type *Help* at the command prompt followed by the first four letters of the command name. For general information on syntax notation, go to *Notation for session commands* in the MINITAB Help index.

Rerunning a Series of Commands

MINITAB generates corresponding session commands for most of the menu commands you used, and stores them in the Project Manager's History folder. Rather than repeat all the previous steps of your analysis using the menus, you can simply rerun these commands by selecting them in the History folder and choosing **Edit ➤ Command Line Editor.**

Session commands for the control chart you just created are stored in the History folder. Use the History folder and the Command Line Editor to re-create the control chart.

Open History folder

1 Choose **Window ➤ Project Manager**.

2 Click the **History** folder.

History folder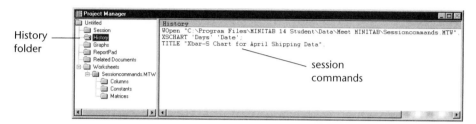

session commands

The right pane of the Project Manager contains all the session commands generated during a MINITAB session. These commands are stored regardless of whether the command prompt is enabled.

When you select any portion of the session commands in the History folder, those commands automatically appear in the Command Line Editor when you open it.

Reexecute a series of commands

1 To highlight the control chart session commands, click XSCHART 'Days' 'Date'; then press (Shift) and click TITLE "Xbar-S Chart for April Shipping Data".

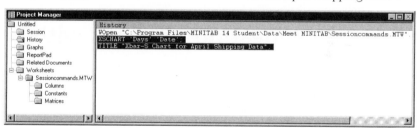

2 Choose **Edit ➤ Command Line Editor**.

3 Click **Submit Commands**.

*Graph
window
output*

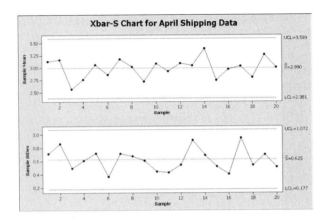

You have created the $\overline{X}$ and S chart in just a few simple steps.

If you edit a graph or a control chart, MINITAB does not automatically generate session commands for the changes made. However, you can generate the session commands, including all editing changes, by using:

- **Editor ➤ Copy Command Language**, which copies the commands to the clipboard.
- **Editor ➤ Duplicate Graph**, which re-creates the graph and stores the session commands in the History folder.

For more information on **Copy Command Language** and **Duplicate Graph**, go to *Copy Command Language (Editor menu)* and *Duplicate Graph (Editor menu)* in the MINITAB Help index.

Repeating Analyses with Execs

An Exec is a text file containing a series of MINITAB commands. To repeat an analysis without using menu commands or typing session commands, save the commands as an Exec and then run the Exec.

The commands stored in the History folder that you used to rerun the above series of commands with the **Command Line Editor** can also be saved as an Exec and executed at any time.

Create an Exec from the History folder

Save the control chart session commands as an Exec. You can use this Exec to continuously analyze the shipping data.

1 Choose **Window ➤ Project Manager**.

2 Click the **History** folder.

3 To select the control chart session commands, click *XSCHART 'Days' 'Date'*; then press ⌂Shift⌃ and click *TITLE "Xbar-S Chart for April Shipping Data"*.

4 Right-click the selected text and choose **Save A**s.

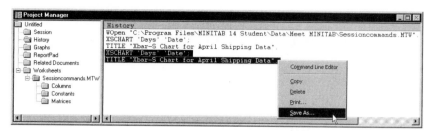

5 In **File name**, type *SHIPPINGGRAPHS*.

6 In **Save as type**, choose **Exec Files (*.MTB)**. Click **Save**.

Reexecute commands

You can repeat this analysis at any time by running the Exec.

1 Choose **File ➤ Other Files ➤ Run an Exec**.

2 Click **Select File**.

3 Select the file SHIPPINGGRAPHS.MTB, then click **Open**.

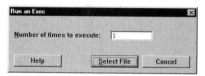

Graph window output

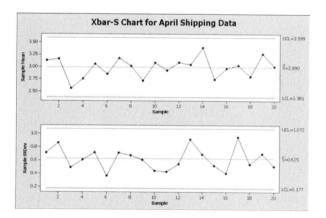

MINITAB executes the commands in the Exec to generate the $\overline{X}$ and S chart. Because you can run the Exec using any worksheet (as long as the column names match), you can share this file with other MINITAB users who need to do the same analysis. For example, the Western shipping center may want to share the control chart Exec with the Central and Eastern shipping centers so they can conduct the same analysis on their own shipping data. If you want to use the Exec with a different worksheet or with different column, edit the Exec using a text editor such as Notepad.

Save project   Save all of your work in a MINITAB project.

1   Choose **File ➤ Save Project As**.

2   In **File name**, type *MY_SESSIONCOMMANDS.MPJ*.

3   Click **Save**.

What Next

You learned how to use session commands as an alternative to menu commands and as a way to quickly rerun an analysis. In the next chapter, you create a report to show the results of your analysis to your colleagues.

6

Generating a Report

Objectives

In this chapter, you:

- Add a graph to the ReportPad, page 6-2

- Add Session window output to the ReportPad, page 6-3

- Edit in the ReportPad, page 6-5

- Save and view a report, page 6-6

- Copy the ReportPad contents to a word processor, page 6-7

- Edit a MINITAB graph in another application, page 6-8

Overview

MINITAB has several tools to help you create reports:

- ReportPad in the Project Manager, to which you can add MINITAB-generated results throughout your sessions

- Copy to Word Processor, which enables you to easily copy content from the ReportPad to a word processor

- Embedded Graph Editor, for editing graphs with MINITAB after you have copied them to other applications

To show your colleagues the shipping data analysis results, you want to prepare a report that includes various elements from your MINITAB sessions.

Using the ReportPad

Throughout *Meet MINITAB*, you performed several analyses and you want to share the results with colleagues. MINITAB's Project Manager contains a folder, called the ReportPad, in which you can create simple reports.

The ReportPad acts as a simple text editor (like Notepad), from which you can quickly print or save in RTF (rich text) or HTML (Web) format. In ReportPad, you can:

- Store MINITAB results and graphs in a single document
- Add comments and headings
- Rearrange your output
- Change font sizes
- Print entire output from an analysis
- Create Web-ready reports

Add graph to ReportPad

You can add components to ReportPad by right-clicking on a graph or Session window output, then choosing **Append to Report**. In addition, text and graphs from other applications can be copied and pasted into MINITAB's ReportPad.

Add the histogram with fits and groups you created in Chapter 2, *Graphing Data*, to the ReportPad.

1 If continuing from the previous chapter, choose **File ➤ New**, then choose **Minitab Project**. Click **OK**. Otherwise, just start MINITAB.

2 Choose **File ➤ Open Project**.

3 Double-click Meet MINITAB, then choose REPORTS.MPJ. Click **Open**.

4 Choose **Window ➤ Histogram of Days**.

5 Right-click anywhere in the graph region, then choose **Append Graph to Report**.

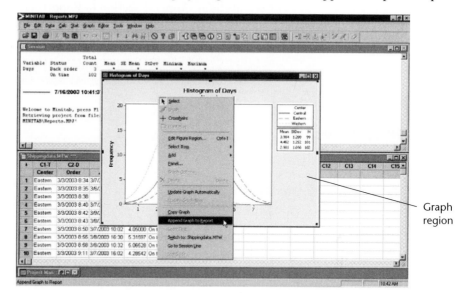

Graph
region

6 Choose **Window ➤ Project Manager**.

7 Click the **ReportPad** folder. The histogram has been added to the ReportPad.

Add Session window output to ReportPad

You also can add Session window output to the ReportPad. In Chapter 3, *Analyzing Data*, you displayed descriptive statistics for the three regional shipping centers. Add the output for the three centers to the ReportPad.

1 Choose **Window ➤ Session**.

2 In the Session window, right-click in the section under the title *Results for Center = Central* and choose **Append Section to Report**. The section of output MINITAB appends is delineated by the output titles (which are in bold text).

If you right-click in this area and choose **Append Section to Report**, the results for the Central shipping center are added to the ReportPad.

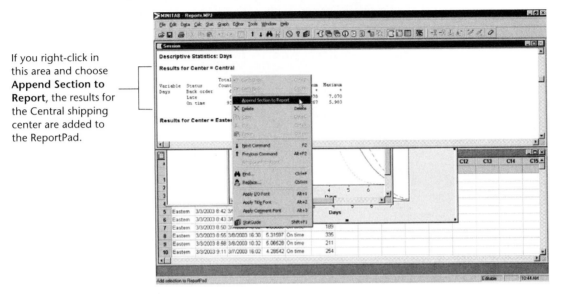

3 Repeat the steps above for the *Results for Center = Eastern* and *Results for Center = Western*.

4 Choose **Window ➤ Project Manager**, then click the **ReportPad** folder. Click to maximize the window to see more of your report.

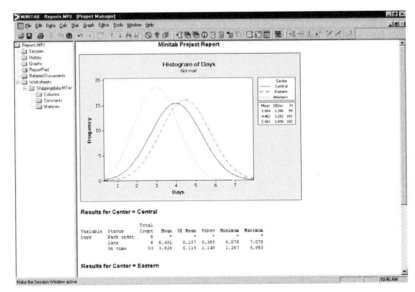

To simultaneously add multiple sections of Session window output to the ReportPad:
1   Highlight the Session window output.
2   Right-click in the Session window.
3   Choose **Append Selected Lines to Report**.

Edit in ReportPad

Customize the report by replacing the default title and adding a short comment to the graphical output.

1   Highlight the default title (**Minitab Project Report**). Type *Report on Shipping Data*. Press ⎡Enter⎤.

2   Below *Report on Shipping Data*, type *Histogram of delivery time by center.*

3   Highlight the text *Histogram of delivery time by center.* Right-click the highlighted text and choose **Font**.

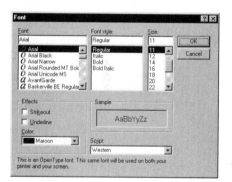

4   From **Font**, choose **Arial**. From **Font style**, choose **Regular**. From **Size**, choose **11**. From **Color**, choose **Maroon**.

5   Click **OK**.

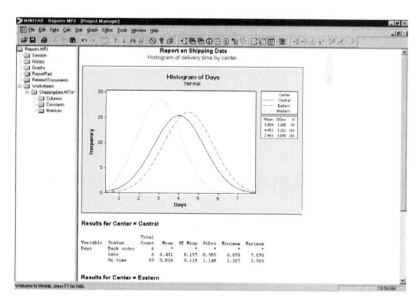

You now have a simple report that illustrates some of your results. If you save a MINITAB project, you can add additional comments and formatting at any time because MINITAB saves the ReportPad contents as part of the project.

All graphs and Session window output remain fully editable after they are appended to the ReportPad. To edit a graph in the ReportPad, double-click the graph to activate MINITAB's embedded graph editing tools.

Saving a Report

You can save the contents of the ReportPad (as well as Session window output and worksheets) either as Rich Text Format (RTF) or Web Page (HTML) so you can open them in other applications.

Save as RTF file

Save your report as a RTF file to send electronically to colleagues or to open in other applications.

1 In the Project Manager, right-click the **ReportPad** folder and choose **Save Report As**.

2 In **File name**, type *ShippingReport*.

3 In **Save as type**, choose **Rich Text Format (*.RTF)**. Click **Save**.

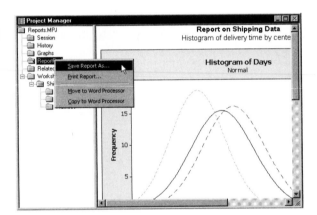

Copying a Report to a Word Processor

Word processors provide formatting options not available in ReportPad, such as adding callouts to highlight important findings and laying graphs side by side.

Two tools in ReportPad, Copy to Word Processor and Move to Word Processor, enable you to transfer the contents of the ReportPad to your word processor without copying and pasting:

■ Copy to Word Processor transfers the ReportPad contents into a word processor while leaving the original contents in the ReportPad.

- Move to Word Processor transfers the ReportPad contents to a word processor and deletes the contents of the ReportPad.

Copy report to a word processor

1 In the Project Manager, right-click the **ReportPad** folder.

2 Choose **Copy to Word Processor**.

3 In **File name**, type *Shipping Report*. You do not need to choose a file type, because Rich Text Format (∗.RTF) is the only option available.

4 Click **Save**.

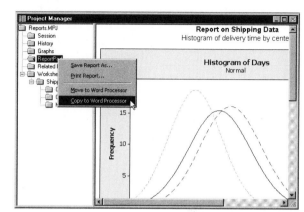

MINITAB automatically opens your default word processor and loads the RTF file you just saved.

You can now edit your MINITAB content in the word processor.

Using Embedded Graph Editing Tools

When you copy graphs to a word processor or other application, either with copy/paste or with Copy to Word Processor, you can use the Embedded Graph Editor to access all MINITAB's graph editing tools.

Edit MINITAB graph in a word processor

To blend the graph into the report background and create a better visual effect, use the Embedded Graph Editor tools to change the fill pattern, borders, and fill lines of the graph without returning to MINITAB.

1   In the word processor, double-click the histogram. Notice that you now have
    several toolbars with editing tools.

MINITAB graph
editing tools

Graph
region

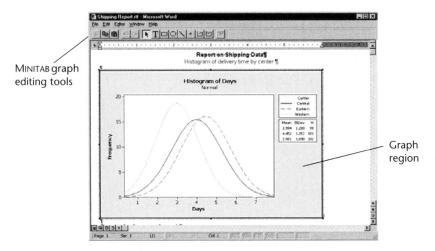

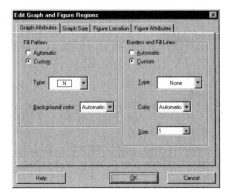

The graph is in edit mode; you can double-click a graph item to edit it as you
would in MINITAB.

2   Double-click in the graph region of the histogram.

3   Under **Fill Pattern**, choose **Custom**.

4   From **Type**, choose .

5   Under **Borders and Fill Lines**, choose
    Custom.

6   From **Type**, choose **None**. Click **OK**.

7   Click outside of the graph to end edit
    mode.

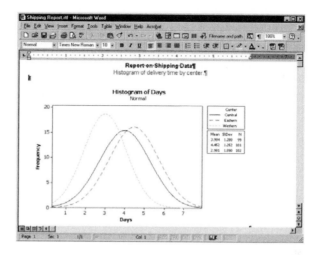

 For more information about MINITAB's Embedded Graph Editor, go to *Embedded Graph Editor* in the MINITAB Help index.

Save project   Save all of your work in a MINITAB project.

1 In MINITAB, Choose **File ➤ Save Project As**.

2 In **File name**, type *MY_REPORTS.MPJ*.

3 Click **Save**.

What Next

In the next chapter, you learn to prepare a MINITAB worksheet. You combine data from multiple sources and place them in MINITAB. Also, to prepare the data and simplify the analysis, you edit the data and reorganize columns and rows.

7

Preparing a Worksheet

Objectives

In this chapter, you:

- Open a worksheet, page 7-2
- Merge data from an Excel spreadsheet into a MINITAB worksheet, page 7-3
- Copy and paste data from a text file into a worksheet, page 7-4
- View worksheet information, page 7-5
- Replace missing value, page 7-6
- Stack columns of data, page 7-6
- Code data, page 7-8
- Add column names, page 7-8
- Insert and name a new data column, page 7-9
- Use the Calculator to create a new worksheet column, page 7-9

Overview

In many cases, you use worksheets that were already set up for you, as you have throughout *Meet MINITAB*. Sometimes, however, you must combine data from different sources and place them in a MINITAB worksheet before beginning an analysis. MINITAB can use data from:

- Previously saved MINITAB worksheet files
- Text files

■ Microsoft Excel documents

To place these data in MINITAB, you can:

■ Type directly into MINITAB

■ Copy and paste from other applications

■ Open from a variety of file types, including Excel or text files

After your data are in MINITAB, you may need to edit cells and reorganize columns and rows to get the data ready for analysis. Common manipulations include stacking, subsetting, specifying column names, and editing data values.

This chapter shows how to place data from different sources into MINITAB and how SHIPPINGDATA.MTW, used in chapters 2 and 3, was prepared for analysis.

Getting Data from Different Sources

For the initial *Meet MINITAB* analyses, the worksheet SHIPPINGDATA.MTW, which contains data from three shipping centers, was already set up. However, the three shipping centers originally stored the book order data in different ways:

■ Eastern—in a MINITAB worksheet

■ Central—in a Microsoft Excel file

■ Western—in a text file

To analyze all the book order data, you must combine the data from all three shipping centers into a single MINITAB worksheet.

Open a worksheet

Begin with data from the Eastern shipping center, which are stored in a MINITAB worksheet called CENTER_EAST.MTW.

1 If continuing from the previous chapter, choose **File ➤ New**, then choose **Minitab Project** and click **OK**. Otherwise, just start MINITAB.

2 Choose **File ➤ Open Worksheet**.

3 Double-click Meet MINITAB, then choose EASTERN.MTW. Click **Open**.

MINITAB can open a variety of file types. To see the file types, click **Files of Type** in the Open Worksheet dialog box.

Merge data from Excel

The Central shipping center stored data in an Excel spreadsheet.

To combine the Central book order data with the Eastern data, merge the data in the Excel spreadsheet with the data in the current MINITAB worksheet.

1 Choose **File ➤ Open Worksheet**.

2 From **Files of type**, choose **Excel (*.xls)**.

3 Choose CENTRAL.XLS.

4 Choose **Merge**.

5 Click **Open**.

Examine worksheet

Choosing **Merge** adds the Excel data to your current worksheet. MINITAB places the data in cells to the right of the current worksheet data in columns C5-C8. If you had not chosen **Merge**, MINITAB would have placed the data in a separate worksheet.

Original data   Merged data

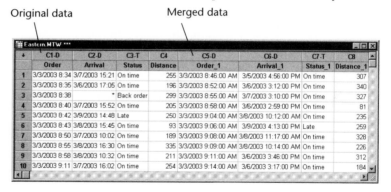

 MINITAB also provides a Merge Worksheets menu command that provides additional options to merge two or more open worksheets. For more information on Merge, go to *Merge Worksheets* in the MINITAB Help index.

Copy and paste from a text file

Instead of opening files that contain data, you can copy data from other applications and paste them into MINITAB. The Western shipping center stored data in a simple text file that you can open using Notepad or WordPad.

1 Open WESTERN.TXT (from the Meet MINITAB folder) in Notepad or any other text editing program.

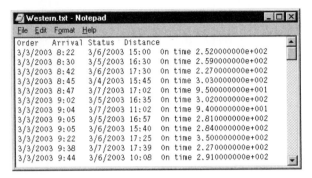

2 Choose **Edit ➤ Select All**.

3 Choose **Edit ➤ Copy**.

4 In MINITAB, click in the column name cell of the first empty column (C9).

If copying and pasting data that includes column names, click in the column name cell first blank column then paste the data. If the data do not include column name cells, click in the first blank cell before pasting the data.

5 Choose **Edit ➤ Paste Cells**.

Examine worksheet

MINITAB pastes the data into the worksheet and fills the appropriate cells in columns C9–C12. The format of this text file was previously set up so that MINITAB would interpret it properly, using the text titles to fill the column name cells and all subsequent data to fill the columns below.

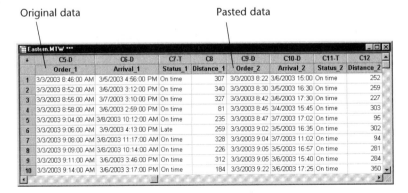

You can also open text files directly in MINITAB using **File ➤ Open Worksheet**. To see what that data will look like in MINITAB, choose **Preview** in the Open Worksheet dialog box.

 MINITAB easily read the above tab-delimited file, but all text files are not in a format that can be easily imported. MINITAB provides several tools for interpreting text file formats. For more information, go to *Text files* in the MINITAB Help index.

Preparing the Worksheet for Analysis

With the data in a single worksheet, you are almost ready to begin the analysis. However, you must modify the worksheet by:

- Replacing a missing value
- Stacking data
- Replacing data
- Adding column names
- Adding a new column
- Creating a column of calculated values

 For a complete list of data manipulations available in MINITAB, go to *Data menu* in the MINITAB Help index.

Show worksheet information

To view a summary of your worksheet columns, use ⓘ on the Project Manager toolbar. This button will open the Project Manager's Columns subfolder in the Worksheets folder. This summary is especially useful in identifying unequal column lengths or columns that contain missing values.

1 Click ⓘ on the Project Manager toolbar or press Ctrl + Alt + I .

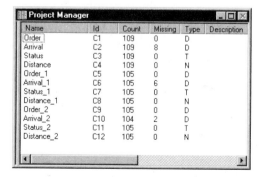

Name	Id	Count	Missing	Type	Description
Order	C1	109	0	D	
Arrival	C2	109	8	D	
Status	C3	109	0	T	
Distance	C4	109	0	N	
Order_1	C5	105	0	D	
Arrival_1	C6	105	6	D	
Status_1	C7	105	0	T	
Distance_1	C8	105	0	N	
Order_2	C9	105	0	D	
Arrival_2	C10	104	2	D	
Status_2	C11	105	0	T	
Distance_2	C12	105	0	N	

The Columns subfolder contains details on the current worksheet. Within each center, the count should be the same for all columns. Notice that the counts for the Eastern data (C1–C4) are 109 for all columns, and the counts for the Central data (C5–C8) are 105 for all columns. However, for the Western center, C10 has a count of 104 unlike the other columns, which have a count of 105.

2 Click again to return to your previous view.

 For more information on the Project Manager toolbar, go to *Project Manager Toolbar* in the MINITAB Help index.

Examine worksheet

Examine C10 to see what value is missing. Notice that the last row of the column is empty. When you copy and paste data from a text or Excel file into a worksheet, MINITAB interprets empty numeric or

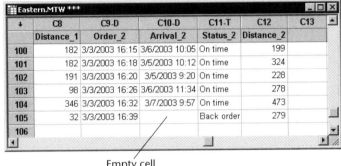

Empty cell

data/time cells as missing values, which appear as asterisks (*) by default. However, if the last row of a column of data in a text file contains an empty cell, MINITAB leaves the cell empty when you paste the data into the worksheet, as you can see in column C10.

Replace missing value

For MINITAB to perform the correct analysis, you must type the missing value symbol in the empty cell of the last row.

1 Click the Data window to make it active, then choose **Editor ➤ Go To**.

2 In **Enter column number or name**, type *C10*.

3 In **Enter row number**, type *105*. Click **OK**.

4 In row 105 of column C10, type an asterisk (*). Press (Enter).

Stack data

Now that the data are assembled in a single MINITAB worksheet, notice the similar variables for each shipping center. Some MINITAB commands allow data from different groups to remain unstacked in separate columns. Others require groups to be stacked, with a column of group levels. However, all analyses can be performed with stacked data.

To analyze the data, you need to rearrange these variables into stacked columns. You can move data within the worksheet by copying and pasting or use Data menu items to rearrange blocks of data.

1 Choose **Data ➤ Stack ➤ Blocks of Columns**.

2 From the list of variables, highlight *Order, Arrival, Status,* and *Distance*. Click **Select** to move the variables into the first row of **Stack two or more blocks of columns on top of each other**.

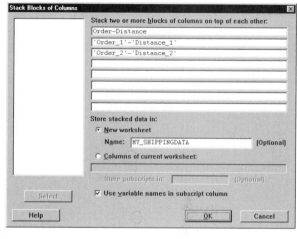

3 Repeat step 2 for the order, arrival, status, and distance columns for the Central and Western shipping centers.

4 Under **Store stacked data in**, choose **New worksheet**. In **Name**, type *MY_SHIPPINGDATA.*

5 Check **Use variable names in subscript column**.

6 Click **OK**.

Examine worksheet

The variables for the shipping centers are all in the same columns, with Order (Eastern center), Order_1 (Central center), and Order_2 (Western center) acting as labels or subscripts to indicate from which shipping center the data originated.

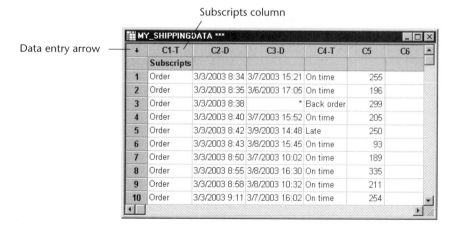

Code data   The labels in the Center column do not adequately indicate from which center the data are from. Code the labels with more meaningful names.

1 Choose **Data ➤ Code ➤ Text to Text**.

2 In **Code data from columns**, enter *Subscripts*.

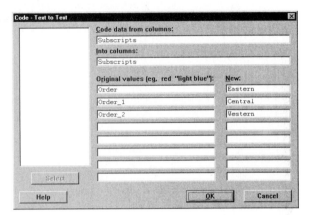

3 In **Into columns**, enter *Subscripts*.

4 In the first row under **Original values**, type *Order*. In the first row under **New**, type *Eastern*.

5 In the second row under **Original values**, type *Order_1*. In the second row under **New**, type *Central*.

6 In the third row under **Original values**, type *Order_2*. In the third row under **New**, type *Western*.

7 Click **OK**.

The shipping center labels in the subscripts column are now Eastern, Central, and Western.

Add column names   Add column names to the stacked data.

1 Click the data entry arrow in the upper left corner of the Data window to make it point to the right.

2 Click in the name cell of **C1**. To replace the label *Subscripts*, type *Center*, then press ⌷Enter⌷.

3 Repeat for the rest of the names:

- In **C2**, type *Order*.

- In **C3**, type *Arrival*.

- In **C4**, type *Status*.

- In **C5**, type *Distance*.

Calculate difference values   Before saving your new worksheet and performing analyses, you need to calculate the number of days that elapsed between order and delivery dates. You can use MINITAB's Calculator to create a column with these values.

Insert and name a column

Insert a column named *Days* between *Arrival* and *Status*.

1 Right-click in C4 and choose **Insert Columns**.

2 Click in the name cell of C4. Type *Days*, then press [Enter].

Use Calculator

Use MINITAB's Calculator to perform basic arithmetic or mathematical functions. MINITAB stores the results in a column or constant.

Compute the delivery time and store the values in the Days column.

1 Choose **Calc ➤ Calculator**.

2 In **Store result in variable**, enter *Days*.

3 In **Expression**, enter *Arrival – Order*. Click **OK**.

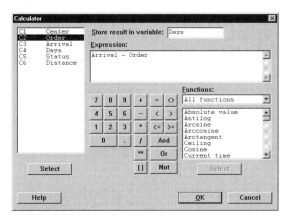

 For more information on MINITAB's Calculator and the available operations and functions, go to *Calculator* in the MINITAB Help index.

Examine worksheet

The Days column contains the newly calculated values that represent delivery time. These values are expressed in numbers of days.

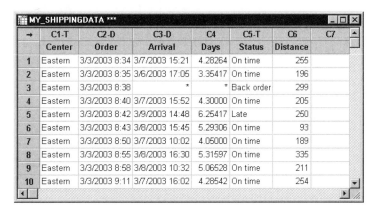

Save worksheet

Save all of your work in a MINITAB worksheet.

1 Choose **File ➤ Save Current Worksheet As**.

2 In **File name**, type *MY_SHIPPINGDATA*.

3 From **Save as type**, choose **Minitab**.

4 Click **Save**.

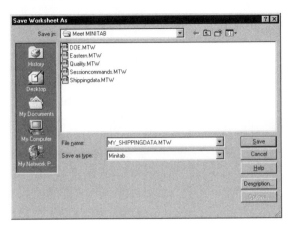

What Next

The shipping center data from several sources are in MINITAB and are set up properly for analysis. In the next chapter, you adjust the MINITAB defaults to expedite future shipping data analyses.

8

Customizing MINITAB

Objectives

In this chapter, you:

- Change default options for graphs, page 8-2
- Create a new toolbar, page 8-3
- Add commands to a custom toolbar, page 8-4
- Assign shortcut keys for a menu command, page 8-5
- Restore your MINITAB default settings using Manage Profiles, page 8-7

Overview

MINITAB has several tools for changing default options or creating custom tools such as individualized toolbars or keyboard shortcuts.

Use **Tools ➤ Options** to change defaults for:

- Program settings (memory usage, initial directory, window layout, and dialog box)
- Data and Session windows
- Statistical commands
- Graphs

Use **Tools ➤ Customize** to:

- Assign a shortcut key to a menu item
- Set options for how MINITAB displays toolbars
- Create custom icons for menu items or toolbar buttons

Now that you have completed your first book shipment analysis and generated a report, you decide to use **Tools ➤ Options** and **Tools ➤ Customize** to tailor the MINITAB environment to make future analyses quicker and easier.

Setting Options

You can change many options during a MINITAB session, such as changing graph display settings or enabling the session command prompt. However, when you exit MINITAB, these options revert back to the defaults for future MINITAB sessions.

If you want a setting to be your default for all MINITAB sessions, use **Tools ➤ Options**. Settings that you change remain active until you change them again.

Because you are planning to do similar analyses on the shipping data during the next few months, you want to change your default preferences.

 If you change options, you can restore MINITAB's default settings at any time. For more information, see *Restoring MINITAB's Default Settings* on page 8-6.

Add automatic footnote

Because you will create the same graphs with similar data in the future, you need a way to distinguish the results of each analysis. You decide to add an automatic footnote to your graphs to include the worksheet name, last modification date, and some information on the data used.

1   If continuing from the previous chapter, choose **File ➤ New**, then choose **Minitab Project** and click **OK**. Otherwise, just start MINITAB.

2   Choose **File ➤ Open Worksheet**.

3   Double-click Meet MINITAB, then choose SHIPPINGDATA.MTW. Click **Open**.

4   Choose **Tools ➤ Options ➤ Graphics ➤ Annotation ➤ My Footnote**.

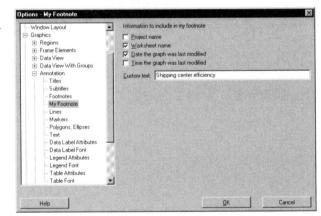

5   Under **Information to include in my footnote**, check **Worksheet name** and **Date the graph was last modified**.

6   In **Custom text**, type *Shipping center efficiency*. Click **OK**.

With these settings, every time you create a graph, MINITAB adds the automatic footnote.

Create a histogram to view footnote

To see an example of the automatic footnote, create a histogram.

1 Choose **Graph ➤ Histogram**.

2 Choose **With Fit and Groups**, then click **OK**.

3 In **Graph variables**, enter *Days*.

4 In **Categorical variables for grouping (0-3)**, enter *Center*.

5 Click **OK**.

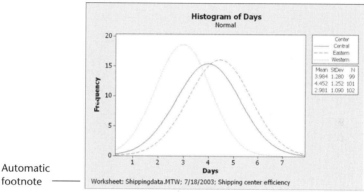

Automatic footnote ———

Creating a Custom Toolbar

In addition to saving time by changing the default options settings for individual commands, you also can save time in future MINITAB sessions by using **Tools ➤ Customize**.

Use **Customize** to create new menus and toolbars that contain only the commands you choose to add, and to assign keyboard shortcuts to commands that you access frequently.

Create a toolbar

During some analyses, you return to the same menu items many times. Combining these items on a single custom toolbar can simplify future analysis.

Create a custom toolbar that includes some of the commands used in the shipping center analysis.

1 Choose **Tools ➤ Customize**.

2 Click the **Toolbars** tab.

3 Click **New**.

4 In **Toolbar Name**, type *Shipping Data*. Click **OK**.

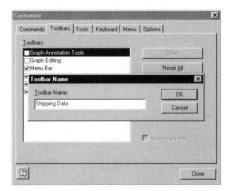

A new blank toolbar labeled *Shipping Data* appears under **Toolbars**, and the new toolbar name appears in the toolbar list.

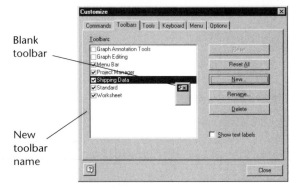

Add commands to the toolbar

Add commands to the blank toolbar. In the shipping center analysis, you used **Graph ➤ Histogram** and **Graph ➤ Scatterplot**, so you want to add these commands to a toolbar.

1 Click and drag the blank toolbar off the Customize dialog box.

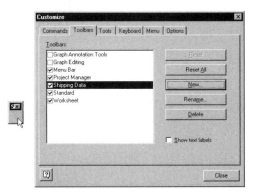

2 Click the **Commands** tab.

3 Under **Categories**, choose **Graph**.

4 Under **Commands**, choose **Histogram**.

Under **Categories** is a list of all MINITAB menus. When you select one of these menus, a list of corresponding menu items appears under **Commands**.

5 Click and drag **Histogram** to the new toolbar.

6 Under **Commands**, choose **Scatterplot**.

7 Click and drag **Scatterplot** to the new toolbar.

8 Click **Close**.

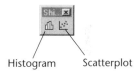

Histogram   Scatterplot

You can add any number of commands until you have a custom toolbar that includes all your frequently used commands. To access the new toolbar items quickly from the keyboard, assign keyboard shortcuts.

You also can create a custom menu. For more information on **Tools ➤ Customize**, go to *Customize* in the MINITAB Help index.

Assigning Shortcut Keys

MINITAB already contains many shortcut keys for frequently used functions such as Copy ([Ctrl]+[C]), Paste ([Ctrl]+[V]), and Save As ([Ctrl]+[S]). Shortcut keys enable you to quickly bypass the menus and open dialog boxes.

To assign shortcut keys, use **Tools ➤ Customize ➤ Keyboard**.

Assign a shortcut key

Because you often create histograms for your shipping data analysis, you want to assign a shortcut key for this command.

1 Choose **Tools ➤ Customize**.

2 Click the **Keyboard** tab.

3 From **Category**, choose **Graph**.

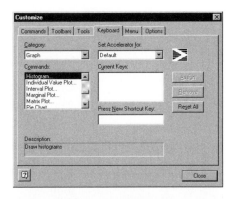

Categories provides a list of all MINITAB menus. When you select one of these menus, a list of corresponding menu items appears under **Commands**.

4 Under **Commands**, choose **Histogram**.

5 Click in **Press New Shortcut Key**.

6 Press ⎡Ctrl⎤+⎡Shift⎤+⎡H⎤.

Under **Press New Shortcut Key**, the **Assigned to** text displays the current status of the selected key combination. In this case, the text reads [**Unassigned**]. Keys or key combinations that are already assigned to a command are indicated here. Any existing combination that conflicts with your choice must be removed from its command before it can be assigned to a new command.

7 Click **Assign**. The new shortcut key appears under **Current Keys**.

8 Click **Close**.

You can now access the Histogram gallery by pressing ⎡Ctrl⎤+⎡Shift⎤+⎡H⎤.

 For a list of MINITAB's default shortcut keys, see the back cover of this book, choose **Help ➤ Keyboard Map**, or go to *Shortcut keys* in the MINITAB Help index.

Restoring MINITAB's Default Settings

Any settings you change using **Tools ➤ Options** and **Tools ➤ Customize**, as well as any changes you have made to date/time data settings or value order settings, are stored in a profile. You can activate and deactivate this profile (and remove all these settings) using **Tools ➤ Manage Profiles**. You also can export and share this profile with other users who are doing a similar analysis.

All settings that you have adjusted while working through *Meet MINITAB* are already stored in your active profile. Deactivate the current profile to restore MINITAB's default settings and change the name of the profile to use for future shipping center analyses.

 For more information on managing profiles, go to *Manage Profiles* in the MINITAB Help index.

Restore defaults

1 Choose **Tools ➤ Manage Profiles**.

2 Click ◄ to move *MyProfile* from **Active profiles** to **Available profiles**.

3 Double-click *MyProfile* in **Available profiles**, then type *ShippingCenterAnalysis*.

4 Click **OK**.

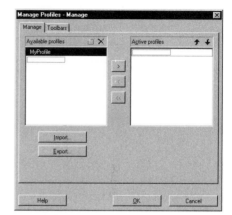

The default settings are now restored. MINITAB creates a new active profile to store any changes you make after this point.

To activate the settings you adjusted during your *Meet MINITAB* sessions, move the current active profile to **Available profiles**, move *ShippingCenterAnalysis* to **Active profiles**, then click **OK**.

 You also can reset MINITAB's defaults by double-clicking the file *RestoreMinitabDefaults.exe* located in the main MINITAB folder installed on your hard drive. Export any profiles you want to keep before running this program.

Save project

Save all of your work in a MINITAB project.

1 Choose **File ➤ Save Project As**.

2 In **File name**, type *My_CUSTOMIZE.MPJ*.

3 Click **Save**.

What Next

Your analysis is complete, but what do you do if you have questions or want more information about a topic? The next chapter suggests ways to get answers to your MINITAB questions and provides details about how to use MINITAB Help and StatGuide.

9

Getting Help

Objectives

In this chapter, you:

- Get answers and find information, page 9-2
- Use MINITAB Help, page 9-6
- Use MINITAB StatGuide, page 9-8
- Use Session Command Help, page 9-10

Overview

If you find yourself with unanswered questions or discover that you need more details about a topic, MINITAB can help.

From assistance with completing a dialog box, to guidance for statistical interpretations, to instructions for using session commands in your analysis, MINITAB's easy-to-use online documentation and Internet resources can help you find the answers you need.

This chapter discusses using Help, StatGuide, and Session Command Help to explore MINITAB and suggests ways to find answers to your MINITAB questions.

Getting Answers and Information

Meet MINITAB focused on only a few of MINITAB's commonly used features. For details about other commands, functions, and statistical concepts, explore MINITAB's documentation and online resources.

Resource	Description	Access
Help	Documentation on MINITAB features and concepts. Includes information on: ■ Menus and dialog boxes ■ Methods and formulas ■ Session commands	■ Click **Help** in any dialog box. ■ Click 🔢 on the toolbar. ■ Press F1 at any time. ■ Choose **Help ➤ Help**. See *Help* on page 9-6 for more information.
How to Use Help	General information on navigating MINITAB Help.	Choose **Help ➤ How to Use Help**.
StatGuide	Statistical guidance that focuses on interpretation of sample results.	■ Right-click in the Session window or a Graph window, then choose **StatGuide**. ■ Right-click in the Session or Graph folder of the Project Manager, then choose **StatGuide**. ■ Click 📲 on the toolbar. ■ Press Shift+F1. ■ Choose **Help ➤ StatGuide**. See *StatGuide* on page 9-8 for more information.
How to Use the StatGuide	General information on using MINITAB StatGuide.	Choose **Help ➤ How to Use the StatGuide**.
Session Command Help	Documentation on MINITAB session commands, which you can use interactively or to create a macro.	■ Choose **Help ➤ Session Command Help**. ■ At the MTB > prompt in the Session window, type *HELP*. ■ To access information on a specific session command, at the MTB > prompt in the Session window, type *HELP* followed by a command. See *Session Command Help* on page 9-10 for more information.

Resource	Description	Access
Tutorials	Step-by-step tutorials that introduce the MINITAB environment and provide an overview of MINITAB.	Choose **Help ➤ Tutorials**.
Meet MINITAB PDF	A PDF version of *Meet MINITAB*. (Adobe Acrobat Reader is provided for your convenience.)	From the Start menu, choose **Programs ➤ MINITAB 14 Student ➤ Meet MINITAB**.
ReadMe	Late-breaking information on this release of MINITAB, including details on changes to the software or documentation.	From the Start menu, choose **Programs ➤ MINITAB 14 Student ➤ ReadMe**.
Technical support	Support is provided only for professors, except for questions about installing the software.	If you are a student with a question about installing the software, use the Minitab Customer Center at *customer.minitab.com* to submit a technical support request. All other questions should be directed to your instructor.
Web site	Learn about our products, training, resources, and more.	Go to *www.minitab.com*.

Please send comments about MINITAB's online and print documentation to doc_comments@minitab.com.

MINITAB Help Overview

The components of MINITAB's online documentation—as well as other related information—are summarized on a single page. From this page, you can proceed to detailed assistance, instructions, and support topics. This overview organizes links to Help topics according to MINITAB's menu structure.

Finding information

To display the overview page:

- Choose **Help ➤ Help**.
- Press F1.
- Click 🕮 on the Standard toolbar.

Three pull-down menus make finding information quick and easy:

- **Basics**—how to use Help, guidelines for getting started with MINITAB, and descriptions of MINITAB windows

- **Reference**—examples of commands, glossary of terms and abbreviations, troubleshooting guidelines, and instructions for using session commands and macros

- **Service and Support**—how to register MINITAB, ways to communicate with Technical Support, and descriptions of MINITAB's documentation, Internet resources, and other products

Use the menus to access
basic facts, reference
material, and services and
support information.

Click a menu link
to view Help topics
for all commands
on that menu.

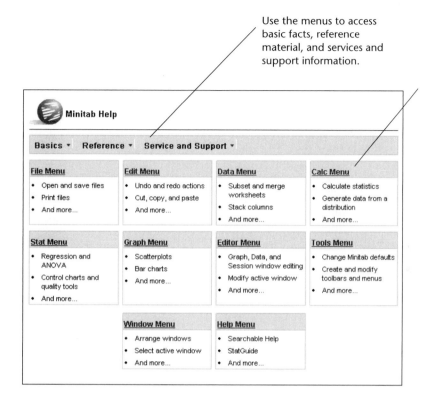

For more information about the MINITAB Help environment, choose **Help ➤ How to Use Help**.

Help

MINITAB Help is a comprehensive, convenient source of information that includes menu and dialog box instructions, overviews, examples, guidance for setting up your data, and methods and formulas. You can explore MINITAB's statistical features and discover new methods for routine tasks. Help also provides guidance on using MINITAB's statistics, quality control, reliability and survival analysis, and design of experiments tools.

Additionally, in Help, you can learn about the MINITAB environment; using session commands; writing macros and Execs; MINITAB's input, output, and data manipulation capabilities; and working with data and graphs.

Finding information

Most Help topics appear in a window that consists of three areas:

- **Toolbar**—contains buttons for hiding and showing the navigation pane, returning to a previous topic, printing one or more topics, and tools for working within the Help environment

- **Navigation pane**—provides four tabs for exploring the table of contents and index, searching for words or phrases, and storing frequently-used topics for quick and easy access

- **Topic pane**—displays the selected Help topic

Toolbar   Topic pane

Navigation pane

- **Contents**—Click any folder or topic for more information.

- **Index**—Search the index for a term, or scroll through the list.

- **Search**—Search the Help for specific words or phrase.

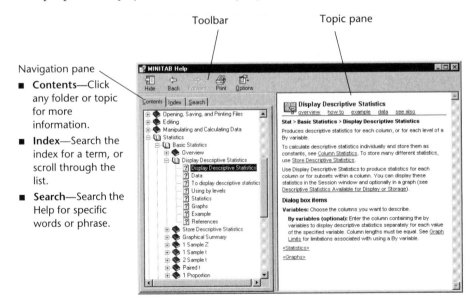

Command-specific information

You can access command-specific assistance from within MINITAB dialog boxes by clicking **Help** in the dialog box or pressing [F1]. Help suggests ways to complete the dialog box and encourages a thorough understanding of the task by supplying links to related topics and associated commands.

Most main dialog box topics contain the following links:

- **Overview** of subject area, including information such as why a certain method is useful and how to choose which method to use

- **How to** instructions on completing the dialog box

- **Example** of using the command, including output and interpretation

- **Data** requirements that explain how you should arrange the data in the worksheet and what data types you can analyze with that command

- **See also** links to related topics and commands, including methods and formulas

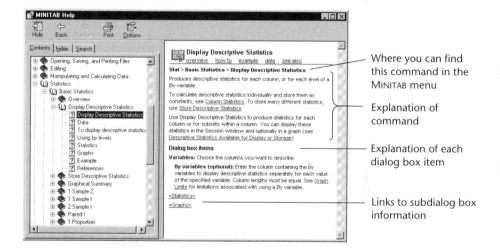

Where you can find this command in the MINITAB menu

Explanation of command

Explanation of each dialog box item

Links to subdialog box information

StatGuide

The MINITAB StatGuide explains how to interpret statistical tables and graphs in a practical, easy-to-understand way. Unlike Help, which provides guidance for using MINITAB, the StatGuide focuses on the interpretation of MINITAB results, using preselected examples to explain the output.

StatGuide topics include information such as:

- Real-life data analysis situations
- Brief summaries of statistical capabilities
- Emphasis on important components of the output

Finding information

After you issue a command, you can learn more about the output by examining StatGuide's sample output and interpretation. The StatGuide provides a direct path to command-specific guidance:

- Right-click in the Session window output or on a graph, then choose **StatGuide**.
- Click in the Session window output or on a graph, then click 🗗 on the toolbar or press Shift+F1.
- In the Project Manager, click the name of the Session window output or graph, then click 🗗 on the toolbar or press Shift+F1. You can also right-click the Session window or graph output name, then choose **StatGuide**.

You can also search the StatGuide to locate specific words or phrases. In MINITAB, choose **Help ➤ Search the StatGuide**, then click the **Find** tab. Or you can access the StatGuide by choosing **Help ➤ StatGuide** and then clicking **Help Topics**.

Click **Help Topics** to open the StatGuide navigation pane.

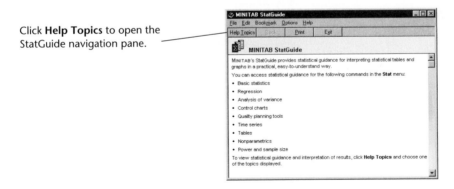

The MINITAB StatGuide navigation pane contains:

■ **Contents**—Explore the StatGuide. The topics appear in the order of MINITAB's Stat menu. Double-click a book to access the menu items.

■ **Index**—Search the index for a term, or scroll through the list of keywords.

■ **Find**—Search the contents of the StatGuide to find all occurrences of a specific word or phrase.

Command-specific information

The StatGuide is arranged in two windows: the main content window and MiniGuide. The MiniGuide contains sample Session window output or graphs and, often, a list of related topics to assist you in navigating the StatGuide. The main content window contains the interpretation of results and links to in-depth content.

Examine sample output and navigate related topics.

Display a list of all StatGuide topics for a command.

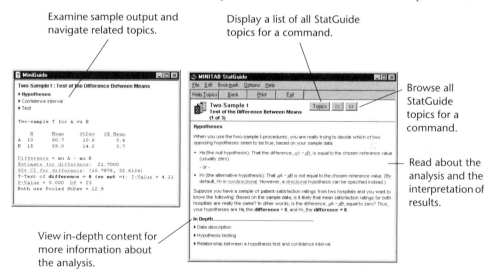

Browse all StatGuide topics for a command.

Read about the analysis and the interpretation of results.

View in-depth content for more information about the analysis.

 For more information about using the StatGuide, choose **Help ➤ How to Use the StatGuide**.

Session Command Help

In addition to using MINITAB's menus and dialog boxes, you can also conduct analyses, generate graphs, and manipulate data using session commands. Each MINITAB menu command has a corresponding session command, which consists of a main command and, usually, one or more subcommands. Session commands are especially useful because they can be used to create macros, which are session commands stored in a text file. Macros can help automate repeated tasks.

Finding information

To access MINITAB Session Command Help, choose **Help ➤ Session Command Help**.

The Session Command Help environment is similar to MINITAB Help. The toolbar, navigation pane, and topic pane provide the necessary tools for learning and using session commands.

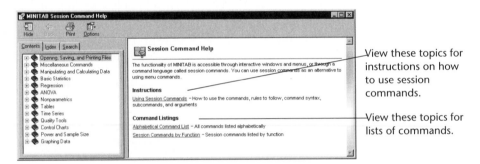

View these topics for instructions on how to use session commands.

View these topics for lists of commands.

 To learn more about session commands, go to Chapter 5, *Using Session Commands.*

**Command-
specific
information**

To access information for a specific session command, at the MTB > command prompt, type *HELP* followed by the command name. Press [Enter].

Most session command topics contain links to:

■ **Example** of using the command, including output.

■ **See also** links to related topics.

Location of the
corresponding command
in the MINITAB menu.

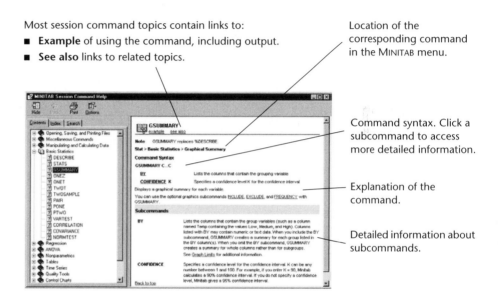

Command syntax. Click a
subcommand to access
more detailed information.

Explanation of the
command.

Detailed information about
subcommands.

What Next

In the next chapter, learn more about the MINITAB environment and the types and forms of data that MINITAB uses. The chapter also includes a list of quick-reference tables of actions and analyses available in MINITAB.

10
Reference

Objectives

In this chapter, you find information about:

- MINITAB environment, page 10-2
- MINITAB data, page 10-5
- Quick reference, page 10-6

Overview

Previous *Meet MINITAB* chapters introduced you to MINITAB and some of its features and commands. This chapter provides in-depth information about the MINITAB environment and data, as well as quick-reference tables to help you to perform the actions and statistics you need in your own analysis.

The MINITAB Environment

As you perform your data analysis, you will work with many different MINITAB windows and tools. Here is a brief overview of the MINITAB environment:

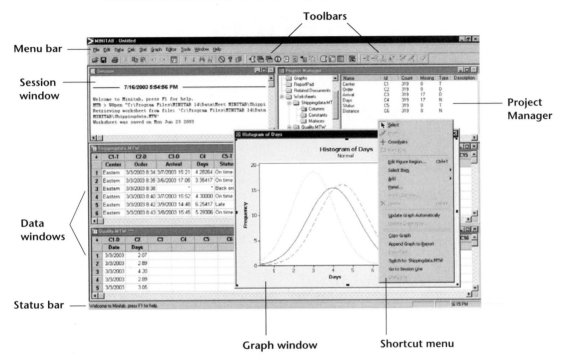

MINITAB windows

- The **Session window** displays text output such as tables of statistics. You can display columns, constants, and matrices in this window by choosing **Data ➤ Display Data**.

- **Data windows** contain columns and rows of cells in which you enter, edit, and view the data for each worksheet.

- **Graph windows** display graphs. You can have up to 200 Graph windows open at a time.

Project Manager

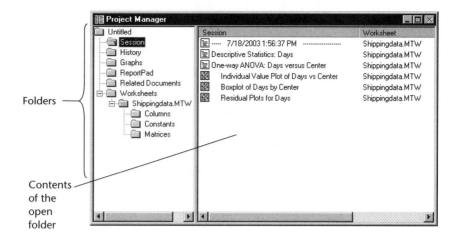

Folders

Contents of the open folder

The Project Manager contains folders that allow you to navigate, view, and manipulate various parts of your project. By right-clicking either the folders or the folder contents, you can access a variety of menus that allow you to manage Session window output, graphs, worksheets, command language, and related project areas.

This folder...	Contains...	Use to...
Session	A list of: ■ All Session window output by command ■ All graphs	Manage Session window output. For example: ■ Jump to Session window output ■ Copy, delete, rename, or print Session window output or graphs ■ Append Session window output or graphs to the ReportPad
History	All commands you have used	■ Repeat complex command sequences ■ Use commands to create Execs and macros
Graph	A list of all graphs in your project	Manage your graphs. For example: ■ Arrange, rename, tile, or remove your graphs ■ Append graphs to the ReportPad

This folder...	Contains...	Use to...
ReportPad	A basic word processing tool	■ Create, arrange, or edit reports of project work ■ Move ReportPad contents to a more powerful word processing program for further editing and layout
Related Documents	A list of program files, documents, or Internet URLs that are related to your MINITAB project	Quickly access project-related, non-MINITAB files for easy reference
Worksheet	The **Columns**, **Constants**, and **Matrices** folders for each open worksheet	View summaries of worksheet information, including: ■ Column counts, missing values, and column descriptions ■ Constants ■ Matrices

Menus and tools

MINITAB provides:

■ A **menu bar** for choosing commands.

■ A **Standard toolbar** with buttons for commonly used functions–the buttons change depending on which MINITAB window is active.

■ A **Project Manager toolbar** with shortcuts to Project Manager folders.

■ A **Worksheet toolbar** with buttons to insert or clear cells, rows, and columns, to move columns, and to move to the next or previous brushed row.

■ A **status bar** which displays explanatory text when you are pointing to a menu item or toolbar button.

■ **Shortcut menus** which appear when you right-click in any MINITAB window or any folder in the Project Manager. The menu displays the most commonly-used functions for that window or folder.

■ Graph editing toolbars (**Graph Editing** and **Graph Annotation Tools**) with buttons for adding and changing graph elements.

 Graph editing toolbars are not visible at start-up, but can be opened by choosing **Tools ➤ Toolbars** and clicking the toolbars you want to show.

Minitab Data

In Minitab, data are contained in a *worksheet*. The number of worksheets a project can have is limited only by your computer's memory.

Data types

A worksheet can contain three types of data:

- *Numeric* data – Numbers.

- *Text* data – Letters, numbers, spaces, and special characters. For example, *Test #4* or *North America*.

- *Date/time* data – Dates (such as Jan-1-2004 or 3/17/04), times (such as 08:25:22 AM), or both (such as 3/17/04 08:25:22 AM). Minitab internally stores dates and times as numbers, but displays them in the format you choose.

Forms of data

Data can be in one of three forms:

Form	Contains...	Referred to by...	Number available
Column	Numeric, text, or date/time data	■ C + number, as in C1 or C22 ■ Column name, such as *Center* or *Arrival*	Limited only by computer memory, up to a maximum of 4000
Stored Constant	A single number or text string (for example, *New York*)	■ K + number, as in K1or K93 ■ Column name, such as *First* or *Counter*	1000
Matrix	A rectangular block of cells containing numbers	■ M + number, as in M1 or M44 ■ Column name, such as *Inverse*	100

The Project Manager Worksheets folder contains a list of the columns, constants, and matrices in each project.

Sample data sets

Minitab comes with a number of sample data sets that are stored in the DATA and STUDNT14 subfolders of the main Minitab folder. The data sets used in *Meet Minitab* are located in the Meet MINITAB subfolder of the DATA folder. For complete descriptions of most of these data sets, go to *Sample data sets* in the Minitab Help index.

Quick Reference

To learn where to find commands in MINITAB's menus, use the quick-reference tables below. Choose a table and scan the first column for the information you need. The second column in each table tells you where to go in MINITAB to perform that action. To find out more about a command, search for the command name in the MINITAB Help index.

The quick-reference tables are:

- Projects, page 10-6
- Worksheets, page 10-7
- Toolbars and menus, page 10-8
- Columns, rows, and cells, page 10-8
- Data manipulation, page 10-10
- Data import and export, page 10-11
- Statistics, page 10-12
- Graphs, page 10-15
- Graph windows, page 10-16

Projects

To...	Choose...
Append Session window output to ReportPad	Window ➤ Project Manager ➤ Session folder, then ReportPad folder
Close current project	File ➤ New ➤ Minitab Project File ➤ Open Project File ➤ Exit
Copy, delete, rename, or print Session window output or graphs	Window ➤ Project Manager ➤ Session folder
Create macros using commands	Window ➤ Project Manager ➤ History folder
Create new project	File ➤ New ➤ Minitab Project
Edit and execute commands used previously	Edit ➤ Command Line Editor
Edit last dialog box	Edit ➤ Edit Last Dialog
Enter or view project description (creator, date, and comments)	File ➤ Project Description
Exit MINITAB	File ➤ Exit

To...	Choose...
Generate, arrange, and edit reports in ReportPad	Window ➤ Project Manager ➤ ReportPad folder
Jump to specific Session window output	Window ➤ Project Manager ➤ Session folder
Manage graphs (save, copy, print, tile, rename, or append to ReportPad)	Window ➤ Project Manager ➤ Graphs folder
Open existing project	File ➤ Open Project
Open project-related, non-MINITAB files, documents, and Internet URLs	Window ➤ Project Manager ➤ Related Documents folder
Repeat complex command sequences	Window ➤ Project Manager ➤ History folder
Run Exec file (type of MINITAB macro)	File ➤ Other Files ➤ Run an Exec
Save project	File ➤ Save Project File ➤ Save Project As
View automatically updated summary of current worksheet	Window ➤ Project Manager ➤ Worksheet folder(s)

Worksheets

To...	Choose...
Change data-entry direction (horizontal or vertical)	Editor ➤ Worksheet ➤ Change Entry Direction
Close worksheet	File ➤ Close Worksheet
Copy worksheet to new worksheet or append to existing worksheet	Data ➤ Copy ➤ Worksheet to Worksheet
Enter or view worksheet description (creator, dates, and comments)	Editor ➤ Worksheet ➤ Description
Merge worksheets	Data ➤ Merge Worksheets
Open existing worksheet	File ➤ Open Worksheet
Open new worksheet	File ➤ New ➤ Minitab Worksheet
Print worksheet	File ➤ Print Worksheet
Save current worksheet with new name	File ➤ Save Current Worksheet As
Save current worksheet	File ➤ Save Current Worksheet

To...	Choose...
Split worksheet	Data ➤ Split Worksheet
Subset all or part of worksheet and copy to a new worksheet	Data ➤ Subset Worksheet

Toolbars and menus

To...	Choose...
Assign keyboard shortcut to command	Tools ➤ Customize, then click Keyboard tab
Create or delete toolbar	Tools ➤ Customize, then click Toolbars tab
Customize menu, submenu, menu bar, or toolbar	Tools ➤ Customize, then click Commands, Toolbars, or Menu tab
Display or hide toolbar	Tools ➤ Customize, then click Toolbars tab
Display toolbar buttons with large icons	Tools ➤ Customize, then click Options tab
Hide or show status bar	Tools ➤ Status Bar
Hide or show toolbar	Tools ➤ Toolbars
Manage user-specific settings	Tools ➤ Manage Profiles
Reset MINITAB menus	Tools ➤ Customize, then click Menu tab
Set options in MINITAB to change defaults to your preferences	Tools ➤ Options
Show or hide screen tips or shortcut keys	Tools ➤ Customize, then click Options tab

Columns, rows, and cells

To...	Choose...
Clear contents of selected cells; leave empty cells or missing-value symbols in their place	Edit ➤ Clear Cells
Combine two or more text columns side by side in new column	Data ➤ Concatenate
Copy columns, constants, and matrices	Data ➤ Copy
Copy contents of selected cells to clipboard	Edit ➤ Copy Cells

To...	Choose...
Cut cells from worksheet and copy to clipboard	Edit ➤ Cut Cells
Delete cells from worksheet, moving up other rows in column	Edit ➤ Delete Cells
Delete rows from the worksheet	Data ➤ Delete Rows
Enter or view column description	Editor ➤ Column ➤ Description
Erase columns, constants, and matrices	Data ➤ Erase Variables
Format columns (data type, width, standard width for all columns in worksheet, hide or unhide, and define value order)	Editor ➤ Format Column Editor ➤ Column
Go to designated cell	Editor ➤ Go To...
Go to next column, active cell, or next/previous brushed row	Editor ➤ Go To ➤ *choose item*
Insert empty cell above selected cell	Editor ➤ Insert Cell
Insert empty column to left of selected column	Editor ➤ Insert Column
Insert empty row above selected row	Editor ➤ Insert Row
Move selected columns to left of designated column, or after last column in use	Editor ➤ Move Columns
Paste contents of clipboard into selected cells	Edit ➤ Paste Cells
Select all cells in worksheet	Edit ➤ Select All Cells
Sort columns and store them in original columns, other columns you specify, or new worksheet	Data ➤ Sort
Stack single columns or blocks of columns	Data ➤ Stack ➤ Columns Data ➤ Stack ➤ Blocks of Columns
Transpose columns into rows and store in new worksheet or at end of current worksheet	Data ➤ Transpose Columns
Unstack columns	Data ➤ Unstack Columns

Data manipulation

To...	Choose...
Calculate column statistics, such as mean, median, or standard deviation	Calc ➤ Column Statistics
Calculate probability densities (pdf), cumulative probabilities (cdf), and inverse cumulative probabilities (invcdf) for chosen distribution	Calc ➤ Probability Distributions
Calculate row statistics, such as mean, median, or standard deviation, for each row of chosen variables	Calc ➤ Row Statistics
Change value or set of values to new values (numeric, text, or date/time data to the same or a different type of data, or use conversion table)	Data ➤ Code
Change data type to/from numeric, text, or date/time	Data ➤ Change Data Type
Define custom lists for Autofill	Editor ➤ Define Custom Lists
Define missing value strings for pasted data	Editor ➤ Clipboard Settings
Display columns, constants, or matrices in Session window	Data ➤ Display Data
Extract one or more parts of date/time data—for example, quarter and year—and put in another column	Data ➤ Extract from Date/Time ➤ To Numeric Data ➤ Extract from Date/Time ➤ To Text
Find/replace data	Editor ➤ Find Editor ➤ Replace
Generate column of ranks for variable	Data ➤ Rank
Generate random data for many distributions, including normal, chi-square, binomial, and Weibull	Calc ➤ Random Data
Make indicator (dummy) variables	Calc ➤ Make Indicator Variables
Make patterned data (simple or arbitrary set of numbers, text values, and simple or arbitrary set of date/time values)	Calc ➤ Make Patterned Data

To...	Choose...
Set starting point for random data generator	Calc ➤ Set Base
Standardize (center and scale) columns of data	Calc ➤ Standardize
Use Calculator to do arithmetic operations, comparison operations, logical operations, functions, and column and row operations	Calc ➤ Calculator
Work with matrices	Calc ➤ Matrices

Data import and export

To...	Choose...
Copy, cut, or paste text in Session window	Edit ➤ Copy Edit ➤ Cut Edit ➤ Paste
Enable or disable command language	Editor ➤ Enable Commands
Find/replace output contents	Editor ➤ Find Editor ➤ Replace
Make output editable or uneditable	Editor ➤ Output Editable
Print Session window	File ➤ Print Session Window
Save Session window output as TXT, RTF, HTM, HTML, or LIS file	File ➤ Save Session Window As
Scroll through output by command	Editor ➤ Next Editor ➤ Previous
Select entire contents of Session window	Edit ➤ Select All
Set fonts to be used in Session window	Editor ➤ Apply Font
View Session window	Window ➤ Session

Statistics

To perform this analysis...	Choose...
Basic statistics	
Calculate column statistics, such as mean, median, or standard deviation	Calc ➤ Column Statistics
Calculate row statistics, such as mean, median, or standard deviation, for each row of chosen variables	Calc ➤ Row Statistics
Descriptive statistics	Stat ➤ Basic Statistics ➤ Display Descriptive Statistics Store Descriptive Statistics Graphical Summary
Z- or t-tests	Stat ➤ Basic Statistics ➤ 1-Sample Z 1-Sample t 2-Sample t Paired t
1 or 2 proportions	Stat ➤ Basic Statistics ➤ 1 Proportion 2 Proportions
Equality of 2 variances	Stat ➤ Basic Statistics ➤ 2 Variances
Correlation or covariance	Stat ➤ Basic Statistics ➤ Correlation Covariance
Normality test	Stat ➤ Basic Statistics ➤ Normality Test
Regression	
Regression (simple/multiple, stepwise, best subsets, or fitted line plot)	Stat ➤ Regression ➤ Regression Stepwise Best Subsets Fitted Line Plot
Logistic regression	Stat ➤ Regression ➤ Binary Logistic Regression

To perform this analysis...	Choose...

ANOVA (analysis of variance)

Analysis of variance	Stat ➤ ANOVA ➤
	One-Way One-Way (Unstacked) Two-Way Balanced ANOVA
Graphical analysis	Stat ➤ ANOVA ➤
	Interval Plot Main Effects Plot Interactions Plot
Test for equal variances	Stat ➤ ANOVA ➤ Test for Equal Variances

Control charts

Variables charts for data in subgroups	Stat ➤ Control Charts ➤ Variables Charts for Subgroups ➤
	Xbar-R Xbar-S Xbar R S
Variables charts for individual data points	Stat ➤ Control Charts ➤ Variables Charts for Individuals ➤
	Individuals Moving Range
Attributes charts	Stat ➤ Control Charts ➤ Attributes Charts ➤
	P NP C U
Time-weighted charts	Stat ➤ Control Charts ➤ Time-Weighted Charts ➤
	Moving Average EWMA

Quality tools

Charts	Stat ➤ Quality Tools ➤
	Run Chart Pareto Chart Cause-and-Effect

To perform this analysis...	Choose...
Time series	
Time series plot	Stat ➤ Time Series ➤ Time Series Plot
Ad hoc model fitting techniques	Stat ➤ Time Series Trend Analysis Decomposition Moving Average Single Exp Smoothing Double Exp Smoothing Winters' Method
Differences and lag	Stat ➤ Time Series ➤ Differences Lag
Correlation analysis	Stat ➤ Time Series ➤ Autocorrelation Partial Autocorrelation Cross Correlation
ARIMA	Stat ➤ Time Series ➤ ARIMA
Tables	
Tally variables	Stat ➤ Tables ➤ Tally Individual Variables
Cross-tabulation and chi-square	Stat ➤ Tables ➤ Cross-Tabulation and Chi-Square
Chi-square test	Stat ➤ Tables ➤ Chi-Square Test (Table in Worksheet)
Descriptive statistics	Stat ➤ Tables ➤ Descriptive Statistics
Nonparametrics	
Median tests	Stat ➤ Nonparametrics ➤ 1-Sample Sign 1-Sample Wilcoxon Mann-Whitney
Analysis of variance by ranks	Stat ➤ Nonparametrics ➤ Kruskal-Wallis Mood's Median Test Friedman
Test of randomness (runs test)	Stat ➤ Nonparametrics ➤ Runs Test

To perform this analysis...	Choose...
Pairwise statistics	Stat ➤ Nonparametrics ➤ Pairwise Averages Pairwise Differences Pairwise Slopes

Power and sample size

Z- and t-tests	Stat ➤ Power and Sample Size ➤ 1-Sample Z 1-Sample t 2-Sample t
1 or 2 proportion	Stat ➤ Power and Sample Size ➤ 1 Proportion 2 Proportions
One-way ANOVA	Stat ➤ Power and Sample Size ➤ One-Way ANOVA

Graphs

To...	Choose...
Examine relationships between pairs of variables	Graph ➤ Scatterplot Matrix Plot Marginal Plot
Examine and compare distributions	Graph ➤ Histogram Dotplot Stem-and-Leaf Probability Plot Empirical CDF Boxplot
Compare summaries or individual values of variables	Graph ➤ Boxplot Interval Plot Individual Value Plot Bar Chart Pie Chart
Assess distributions of counts	Graph ➤ Bar Chart Pie Chart

To...	Choose...
Plot a series of data over time	Graph ➤ Time Series Plot Area Graph Scatterplot
Display character graphs (must be added via Tools ➤ Customize ➤ Menu)	Character Graphs ➤ *choose graph*

Graph windows

To...	Choose...
Add gridlines, reference lines, data labels, titles, or other items to graph	Editor ➤ Add
Add variables to brushing table	Editor ➤ Set ID Variables
Bring selected annotation element to front or send to back	Editor ➤ Annotation ➤ Bring to Front Editor ➤ Annotation ➤ Send to Back
Brush graphs	Editor ➤ Brush
Copy command language of graph, including for editing	Editor ➤ Copy Command Language
Copy graph to paste into another application	Edit ➤ Copy Graph
Copy selected graph text	Editor ➤ Copy Text
Create column that identifies brushed rows	Editor ➤ Create Indicator Variables
Deselect graph element	Editor ➤ Select
Duplicate annotation	Editor ➤ Annotation ➤ Duplicate Annotation
Duplicate graph	Editor ➤ Duplicate Graph
Edit selected element of graph	Editor ➤ Edit *selected element*
Layout different graphs on same page	Editor ➤ Layout Tool
Make similar graph by changing only variables	Editor ➤ Make Similar Graph
Open graph	File ➤ Open Graph
Panel graphs of different groups in same graph window	Editor ➤ Panel

To...	Choose...
Print graph	File ➤ Print Graph
Rotate selected annotation element	Editor ➤ Annotation ➤ Rotate Left *or* Rotate Right
Save graph (MINITAB MGF, JPG, TIF, PNG, or Windows BMP)	File ➤ Save Graph As
Select graph element for editing	Editor ➤ Select Item ➤
Show or hide graph annotation toolbar	Editor ➤ Annotation ➤ Graph Annotation Tools
Update graph when data change	Editor ➤ Update
View exact xy coordinates of point on graphs with standard two-variable regions	Editor ➤ Crosshairs
Zoom in and out on graph	Editor ➤ Zoom

Index

A

adding data to a worksheet 4-5
analysis of variance 3-4
    Tukey's multiple comparison
       test 3-4
analyzing data 3-1
annotating graph layout 2-13
annotation, automatic 8-2
ANOVA
    see analysis of variance
Append to Report 6-2
arithmetic functions
    see Calculator
arrow, data entry 4-5
assessing quality 4-1
Autofill 4-5
automatic footnote, creating 8-2
automating an analysis 5-6

B

boxplots of data 3-5
built-in graphs 2-1, 3-1
    generating 3-5

C

Calculator 7-9
center line 4-2
    interpreting 4-8
changing defaults 8-2
coding data 7-8
columns 1-5, 10-5

inserting 7-9
naming 7-8
number 1-5
stacking 7-6
Command Line Editor 5-4
command prompt 5-2
confidence intervals 3-6
constants 10-5
contacting MINITAB 9-3
control charts 4-2
    adding reference line 4-7
    setting options 4-2
    subgroups 4-3
    updating 4-6
control limit 4-2
Copy to Word Processor 6-6
copying and pasting data 7-4
custom toolbars, creating 8-3
customer support 9-3
customizing MINITAB 8-1

D

data
    adding to a worksheet 4-5
    analyzing 3-1
    coding 7-8
    copying and pasting 7-4
    date/time 10-5
    forms 10-5
    numeric 10-5
    replacing 7-8
    stacking 7-6
    text 10-5
    types 1-5, 10-5
data entry arrow 4-5
data folder, setting default 1-4

data sets, sample 10-5
Data window 1-3
date/time data 10-5
default settings
    changing 8-2
    data folder 1-4
    graphs 2-6
    restoring 8-2, 8-6
descriptive statistics, displaying 3-2
Display Descriptive Statistics 3-2

E

editing graphs 2-5
editing in ReportPad 6-5
editing tools for graphs 6-7
Embedded Graph Editor 6-7
environment, in MINITAB 10-2
Excel, merging data into worksheet
    7-3
Exec file 5-6

F

files
    HTML format 6-6
    merging 7-3
    MPJ file type 2-14
    MTB file type 5-7
    MTW file type 7-2
    opening a worksheet 7-2
    RTF format 6-6, 6-7
    saving projects 2-13
    text 7-4
    types used by MINITAB 7-2
    XLS format 7-3

preparing a worksheet 7-1
previewing a worksheet 7-5
printing 2-13
profiles, managing 8-6
project files, saving 2-13
Project Manager 3-10
    Graph folder 10-3
    History folder 5-5, 10-3
    Info window 7-5
    Related Documents 10-4
    ReportPad 6-2, 10-4
    Session folder 10-3
    Show Graphs icon 3-11
    Show Session Folder icon 3-10
    toolbar 3-10, 10-4
    Worksheet folder 10-4

Q

quality 4-1
quick reference 10-6

R

ReadMe file 9-3
reference line 4-7
Related Documents folder 10-4
repeating an analysis 5-5
replacing values in a worksheet 7-8
ReportPad 6-2, 10-4
    adding graphs 6-2
    adding Session window output 6-3
    changing font 6-5
    editing 6-5
    saving contents 6-6
reports
    copying to word processor 6-6
    saving 6-6
residual plots 3-5
    four-in-one 3-8
    histogram of the residuals 3-7
    normal probability plot 3-7
    residuals versus order 3-7

residuals versus the fitted values 3-7
restoring default settings 8-2, 8-6
rows 1-5
RTF file format 6-6, 6-7

S

sample data sets 10-5
saving
    Execs 5-6
    project 2-13
    report 6-6
    worksheet 7-10
scatterplot 2-9
    editing 2-10
    interpreting 2-10
Session Command Help 9-10
    command-specific information 9-11
    finding information 9-10
session commands 5-1
    enabling 5-2
    generating for edited graph 5-6
    using 5-1
Session folder 10-3
Session window 1-3, 10-2
    adding output to ReportPad 6-3
    command prompt 5-2
    viewing output 3-3
setting options 8-2
shortcut keys
    assigning 8-5
    default 8-6
shortcut menus 10-4
Show Graphs icon 3-11
Show Session Folder icon 3-10
special causes 4-2
stability 4-2
stacking data 7-6
Standard Toolbar 10-4
starting MINITAB 1-3
StatGuide 9-8
    accessing 3-8, 9-8

command-specific information 9-9
    finding information 9-8
status bar 10-2, 10-4
stored constants 10-5
subgroups 4-3
subscripts 7-7

T

technical support 9-3
tests for special causes 4-2
    setting options 4-2
text
    data 10-5
    files 7-4
time data
    see date/time data
toolbars 10-2
    creating custom 8-3
    Graph Annotation Tools 10-4
    Graph Editing 10-4
    Project Manager 3-10, 10-4
    Standard 10-4
    Worksheet 10-4
Tukey's multiple comparison test 3-4
    interpreting 3-6
    StatGuide 3-8
typographical conventions 1-2

U

updating graphs 4-5

V

variables 1-5
    entering in a dialog box 2-3
viewing
    graphs 3-11
    Session window output 3-3

W

X

Documentation

To help you use MINITAB most effectively, Minitab offers a variety of helpful documentation.

Meet MINITAB: *Meet MINITAB* is a concise guide to getting started with MINITAB software.

MINITAB Help: This comprehensive, convenient source of information is available at the touch of a key or a click of the mouse. In addition to complete menu and dialog box documentation, you can find overviews, examples, guidance for setting up your data, information on calculations and methods, and a glossary.

MINITAB StatGuide: The online StatGuide explains how to interpret statistical tables and graphs in a practical, easy-to-understand way. The tone is informal and friendly, and it is easily accessible by right-clicking output or by clicking the icon in the toolbar. From basic statistics, to quality tools, to design of experiments, you'll get easy-to-understand guidance when you need it.

Tutorials: The tutorials help you learn MINITAB quickly. You can find them on the Help menu.

Companion Text List: The Companion Text List, updated continuously, is a resource for statisticians, teachers, and MINITAB users. The CTL is a bibliographical listing of currently available texts that feature MINITAB Statistical Software, including textbooks, textbook supplements, and other related teaching materials. For a complete bibliography, check our Companion Text List at http://www.minitab.com/resources/ctl/.

MINITAB Handbook, Fourth Edition: A supplementary text that teaches basic statistics using MINITAB, the Handbook features the creative use of plots, application of standard statistical methods to real data, in-depth exploration of data, and more. To order, please contact your nearest Minitab office.

We appreciate your feedback! If you find errors or problems within any of MINITAB's documentation, please notify us by e-mailing doc_comments@minitab.com.

Additional MINITAB Products

Minitab offers a collection of software, support materials and services that enable you to manage your quality and process improvement processes. Please contact your nearest Minitab office to receive additional information about the following:

Process management software: Minitab Quality Companion™ enables you to manage and coordinate the "soft" tasks of process improvement–such as process mapping, brainstorming, and consensus building.

Other language products: In our continuing effort to support the global community, Minitab has product and documentation offerings in several languages. Presently, French, German, Japanese, and Korean products are available.

Student product: MINITAB Student Software is a streamlined and economical version of Professional MINITAB, designed specifically for introductory and business statistics courses. It comes bundled with a wide variety of textbooks from leading textbook publishers.

Educational resources: Explore a wealth of additional educational resources at http://www.minitab.com/education.

Training: Without question, MINITAB is one of the easiest statistics packages to use, but to fully maximize its power you'll want to take advantage of the wide variety of training courses that we offer. Sessions are available for beginning to advanced users, and are tailored to address the specific needs of various industries. You can find more information at www.minitab.com/training.

How to Order Additional Products

To order, contact Minitab Inc., Minitab Ltd., Minitab SARL, or your local partner. Contact information is provided on the back cover of this book. Or, visit our web site at www.minitab.com.